WORLD OF WONDER — I KNOW ABOUT!

Children will be WOWed when they dive into this fun, fact-packed series that explores the most fascinating things in the world around them. From the human body, to sharks, to insects, to atlases, to dinosaurs, there is something new and exciting to learn at every turn!

These books are anchored by the philosophy that the best form of learning is through high-quality, multi-faceted experiences. This series provides lasting knowledge as it engages children on many levels. The graphics and illustrations inspire learning in readers of all ages as they explore everything from the depths of the ocean to the cells of the human body. Students will also be captivated by the fun facts and interesting explanations, which bolster a comprehensive understanding of each subject area.

Teachers and parents alike will appreciate the significance of this series as an addition to their libraries, as it promotes self-sufficiency in the learning process. With vividly portrayed information on many of children's favorite subjects, these books promote an enjoyment for both discovering and sharing knowledge, key components of a successful learning environment.

Learning should be fun, interesting, and applicable, and we are sure the young learners in your life will find this series to be all of that – and more!

"You can teach a student a lesson for a day; but if you can teach him to learn by creating curiosity, he will continue the learning process as long as he lives."

– Clay P. Bedford

WORLD OF WOW WONDER
I KNOW ABOUT!

ATLAS 6

40 SHARKS

BODY 68

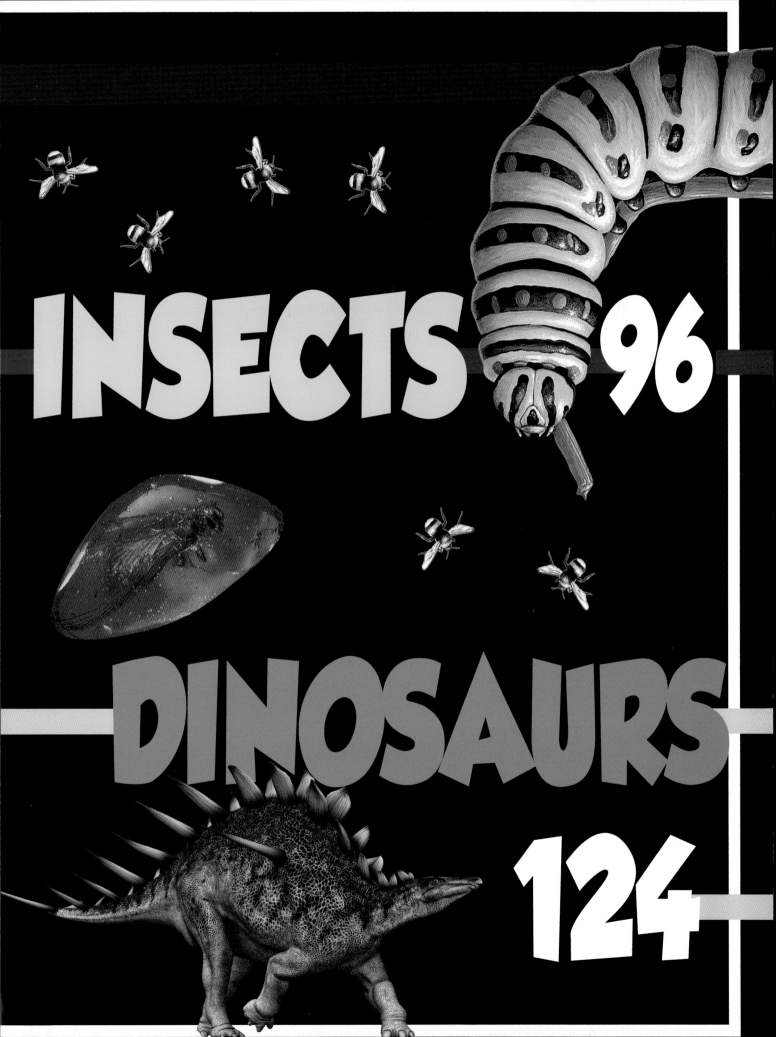

I KNOW ABOUT!

ATLAS

RIVERS, LAKES, AND SWAMPS 14

FORESTS 17

THE MAPS 21

NORTH AMERICA 22

EUROPE 27

AFRICA AND THE MIDDLE EAST 29

NORTHERN ASIA 24

SOUTH AND SOUTHEAST ASIA 33

THE PACIFIC ISLANDS 35

ANTARCTICA 36

FORMATION *of the* EARTH

 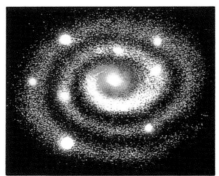 

THE EARTH FORMS

Over 4 billion years ago, the solar system was a swirling disk of gas and dust (above). Because of gravity, the dense ball at its center started to collapse on itself, until it ignited and became the Sun. Meanwhile, the dust and rock in the disk began to clump together to form the planets. Today, the solar system contains 8 planets, over 60 moons, and countless asteroids and comets.

The Earth's changing face

During its earliest days, about 4 billion years ago, the Earth was a hot, waterless place. Its surface was pounded by comets and meteorites (below), while volcanoes released water vapor and other gases.
These gases created an early atmosphere as well as the first seas and oceans. Volcanic activity began to decrease and the planet became a suitable place for life to evolve about 3.5 billion years ago. The earliest supercontinent, called Pangea, appeared about 1.5 billion years ago.

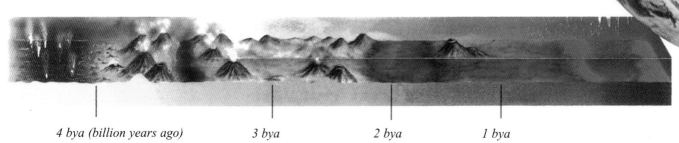

4 bya (billion years ago)   3 bya   2 bya   1 bya

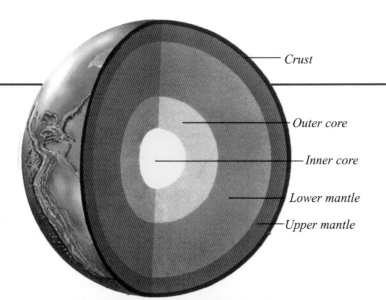

Crust
Outer core
Inner core
Lower mantle
Upper mantle

INSIDE THE EARTH

The outer crust of the planet is a thin shell, about 31 miles (50 km) deep (left). Below this, stretching down 1,875 miles (3,000 km) into the Earth, is the mantle made up of rock that has been turned liquid by the high temperature and pressure. At the very center, about 4,062 miles (6,500 km) below the crust, is the solid core made up of iron.

Blue sphere
Today's Earth appears as a blue sphere, covered with white clouds. This is due to the large amount of water (below).

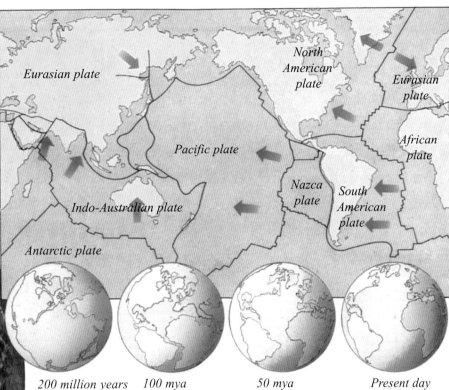

200 million years ago (mya)　　100 mya　　50 mya　　Present day

PLATES

The outer crust is split into about a dozen plates, like an enormous jigsaw puzzle. However, few of these plates are stationary—many of them move about 2 inches (5 cm) every year. Over millions of years, this plate movement has dragged the landmasses apart. About 200 million years ago, this movement broke up the original supercontinent of Pangea ("all lands"). Since then, the landmasses have been split up farther, creating the continents that we have today (above).

Core　　*Convection currents*　　*Mantle*

CONVECTION CURRENTS

Just as smoke rises from a chimney and steam rises from a kettle, liquid rock (called magma), heated by the Earth's core, rises toward the surface. When this magma reaches the crust it cools, it gets pushed aside by more liquid rock rising from below, and sinks back through the mantle. These magma movements are called "convection currents." Geologists believe that these currents pull the plates in the Earth's crust apart and push them together (left).

The MOVING EARTH

EARTHQUAKES AND VOLCANOES

The map (below) shows that most earthquakes and volcanic eruptions occur around plate boundaries—the regions where two or more plates pull apart, crash into one another, or scrape against each other. In some areas away from boundaries the crust may be very thin. In these areas, called "hot spots," liquid rock can force its way through and form volcanic islands, such as the Hawaiian Islands in the Pacific Ocean.

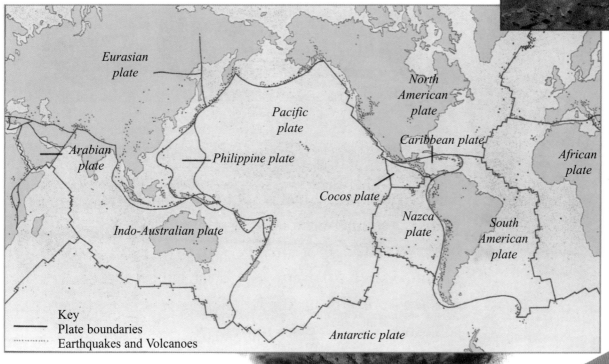

Ring of Fire
The "Ring of Fire" is a chain of volcanoes around the Pacific plate. It includes the volcanoes in the Andes, the western United States, and eastern Asia.

ERUPTION

A volcano occurs where molten rock, or magma, breaks through the Earth's crust. When it reaches the surface the hot rock is called "lava." It can erupt through the vent as a boiling cloud of gas and smoke (above), rivers of glowing liquid rock (far right), or a mixture of the two. After a very violent eruption, the vent of the volcano may collapse to form a caldera.

SHOCKWAVES

When two crust plates scrape against each other, they rarely do so smoothly. Instead, they will move in a series of jerks. When there is a sudden jerk along the fault line (the line where the two plates meet), a huge amount of energy is released causing an earthquake. This energy is sent out in horizontal and vertical shockwaves from the earthquake's source. This source is called the "focus" (right).

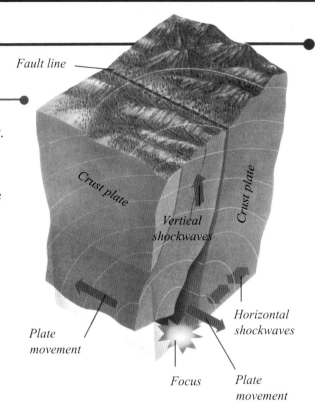

San Andreas Fault
The San Andreas Fault (above) lies between the Pacific and North American plates. The last major earthquake along the fault occured in 1989.

PLATE BOUNDARIES

When two plates move apart, they leave a crack in the earth. Magma forces its way up and cools, forming a ridge running along the crack. When two plates collide, one is sometimes forced underneath. As it is pushed down, it melts, and the molten rock forces its way to the surface to form chains of volcanoes along the boundary.

MOUNTAINS

Where two continents crash into each other, the impact will crumple and force up the rock in between, forming huge chains of mountains. These mountain ranges include the Himalayas in southern Asia and the Alps in Europe.

Growing mountains
The Himalayas (above) are being forced up by the Indo-Australian and the Eurasian plates.

RESOURCES

FOSSIL FUELS

Fossil fuels, such as coal, oil, and gas, are all formed from the remains of long-dead organisms, such as plants and microscopic bacteria. These remains were buried under layers of rock. Over time, the pressure from these layers of rock transformed the dead organisms into fossil fuels that can be converted into fuel for anything from power stations to cars.

The world's largest oil and gas fields are found in the United States, the North Sea (above and right), North Africa, and the Middle East. The largest coal fields are found throughout northern Europe, as well as Siberia, Alaska, and Australia.

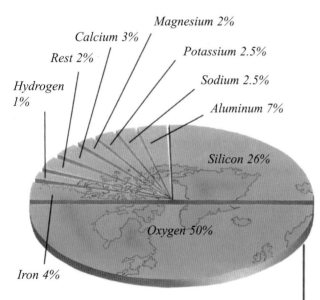

THE EARTH'S CRUST

Hydrogen 1%
Rest 2%
Calcium 3%
Magnesium 2%
Potassium 2.5%
Sodium 2.5%
Aluminum 7%
Silicon 26%
Oxygen 50%
Iron 4%

Together, the elements oxygen and silicon make up 76 percent of the Earth's crust (above). A further 22 percent is formed by six metals—iron, aluminum, calcium, sodium, potassium, and magnesium. The remaining 3 percent is made up from other elements, including hydrogen and carbon.

MINERALS

Minerals such as quartz, copper, and gold are found all over the world. The richest concentrations of minerals include the gold fields in South Africa and the copper belt in Zambia.

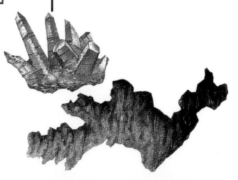

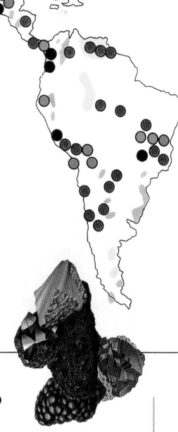

Alaskan coal field

ORES

These are rocks that contain enough of a metal to make it worth mining. Metals commonly found in ores include gold, silver, platinum, tin, iron, and copper. Once they have been mined, these metals have to be removed from the rocks that contain them.

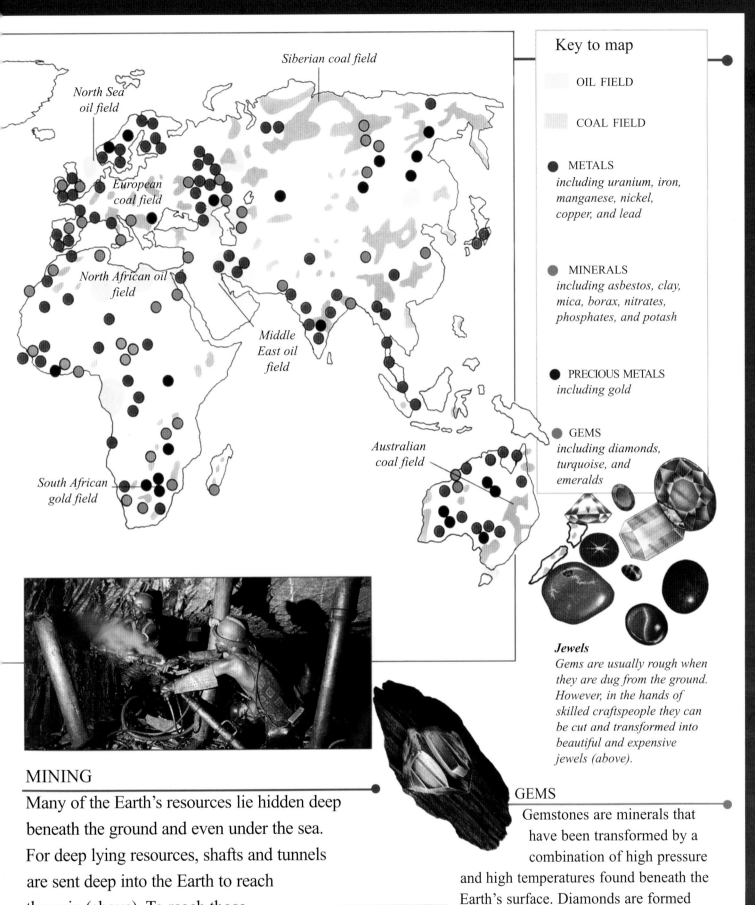

Key to map

	OIL FIELD
	COAL FIELD
●	METALS *including uranium, iron, manganese, nickel, copper, and lead*
●	MINERALS *including asbestos, clay, mica, borax, nitrates, phosphates, and potash*
●	PRECIOUS METALS *including gold*
●	GEMS *including diamonds, turquoise, and emeralds*

Jewels
Gems are usually rough when they are dug from the ground. However, in the hands of skilled craftspeople they can be cut and transformed into beautiful and expensive jewels (above).

MINING

Many of the Earth's resources lie hidden deep beneath the ground and even under the sea. For deep lying resources, shafts and tunnels are sent deep into the Earth to reach the vein (above). To reach those near the surface, strip mining can be used where entire layers of the ground are removed to extract the resource (right).

GEMS

Gemstones are minerals that have been transformed by a combination of high pressure and high temperatures found beneath the Earth's surface. Diamonds are formed from carbon, the same element that makes up graphite. When the carbon is subjected to high pressures and heat deep within the Earth, it crystallizes to form the valuable gem.

RIVERS, LAKES, and SWAMPS

Waterfalls

YOUTH

MATURITY

Meandering river

Delta

OLD AGE

River delta

FRESHWATER

Freshwater makes up only 3% of the world's water. However, nearly all of this water is frozen in glaciers and the ice caps found at the Poles. Only 0.01% of the world's water flows through rivers, lakes, and swamps (below).

FRESHWATER LIFE

Many rivers, swamps, and lakes teem with life from fish (above) to frogs (below). The largest freshwater fish is the pla beuk found in the Mekong River in Southeast Asia.

RIVERS

Rivers form a part of the water cycle. They act as channels, carrying rain or spring water to lakes or an ocean. As they flow, they gouge their way through rock, creating physical features such as gorges. The rock that has been eroded is carried by the river and later deposited to form features such as deltas. Near its source, a river is described as "young" (left). As it nears its destination, whether this is a lake or an ocean, it reaches "maturity," and flows from side to side in meanders. At its destination, it reaches "old age," when most of its load is deposited.

SEAS and OCEANS

WATERY WORLD

The seas and oceans around the world hold 97% of the world's water, and cover over two thirds of the planet's surface. The largest ocean, the Pacific, covers 63,855,000 square miles (165,383,000 sq km). The deepest point in the oceans is the Mariana Trench in the Pacific. It is nearly 7 miles (11 km) deep.

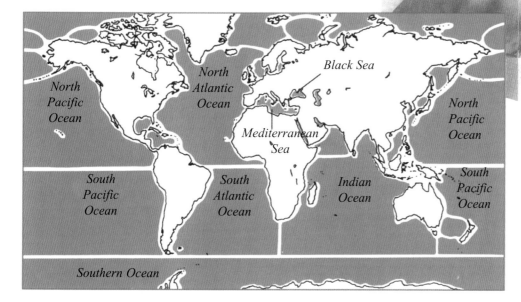

OCEAN FLOOR

Surrounding a landmass is a shallow part of the sea called the "continental shelf." At about 63 miles (100 km) from the shore, the sea floor drops off sharply down the continental slope, before arriving at the ocean floor about 4 miles (6 km) below the surface.

LIFE BENEATH THE WAVES

The seas hold an amazing range of life. Coral reefs (below), found off the coasts of Australia, Central America, and Africa are home to animals ranging from the tiny coral to clams, stingrays, and an array of brightly colored fish. Life in the deep oceans includes microscopic plankton. These tiny organisms form the basic food supply for others, from crabs (above) to enormous whales.

COASTAL EROSION

The powerful forces of the seas and oceans can have devastating effects on the rocks and soil of the coastline, eroding huge amounts from some areas and depositing them in others. The results can be stunning, with such formations as the enormous rock stacks of the Twelve Apostles on Australia's South Coast (above).

MOUNTAINS

JAGGED PEAKS

Mountains and hills are formed as the result of two crust plates colliding, causing the rock in between to be crumpled up into jagged peaks, such as the Himalayas. They may also be the result of volcanic activity, and today, many chains of mountains still have active volcanoes, such as Mount St. Helens in Washington state (below).

Plant altitudes
As the climate changes with altitude on a mountain slope, so different types of plants will grow at different heights (above and below). At the top of the highest peaks are snow and ice where little grows (1). Below this is a strip of Alpine meadow (2), containing some flowering plants and grasses. The highest trees are found in the band of coniferous evergreen forest (3). These are followed by deciduous trees (ones that shed their leaves, 4) and, if the climate is warm and wet enough, tropical cloud forest (5). Finally, there are more bands of deciduous (6) and coniferous (7) trees before giving way to open grassland at the foot of the mountain (8).

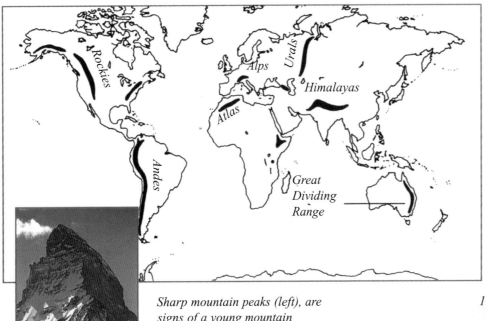

Sharp mountain peaks (left), are signs of a young mountain range that has not been eroded to a great extent.

Smoother mountains (right) indicate an old mountain range. This is because they have been eroded to their present shape.

WILDLIFE

Mountain wildlife has to cope with harsh environments. Few large and ornate plants can survive the unpleasant conditions. Instead, the plants tend to be small, such as the mountain avens (above left). Mountain animals must adapt to survive. The big horn sheep of the Rocky Mountains have a thick fleece and are sure-footed climbers (right).

FORESTS

WOODS AND FORESTS

Large areas of land on almost every continent, with the exception of Antarctica, are covered in forests of one kind or another. These can be the enormous boreal forests (see below) that stretch across the northern regions of North America, Asia, and Europe, or the concentrations of rainforest that fill the centers of South America and Africa. These great areas of land are under threat from massive deforestation programs that clear forests at an alarming rate—up to 54,687 square miles (142,000 sq km) each year.

BOREAL FORESTS

These are found in areas with extremely cold winters and a short growing season. Also called "taiga," they generally consist of one type of tree—coniferous evergreens, such as pines (above). Despite the often harsh conditions, some animals, including the pine marten (right), thrive in boreal forests.

The world's woodlands
The large region of woodland stretching across Russia and Siberia is the largest forest on the planet. It ranges for over 6,250 miles (10,000 km) from the Baltic Sea in the east to the Pacific Ocean in the west.

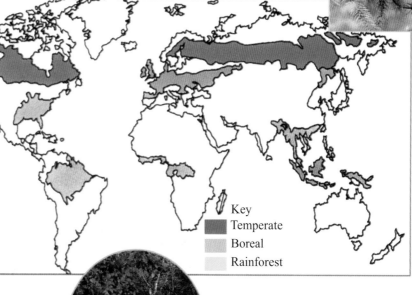

Key
Temperate
Boreal
Rainforest

TEMPERATE FORESTS

These can consist of either deciduous (trees that lose their leaves) or evergreen trees. They grow in areas with warm summers, cool winters, and plenty of rainfall. They are found in areas such as Central Europe and the South. They are able to support an immense amount of wildlife, including squirrels (right) and deer.

RAINFORESTS

One of the most diverse and richly populated forms of environment, these forests are found in the warmer and wetter parts of the world, such as South America, Central Africa, and Southeast Asia. The rich growing conditions ensure the growth of a huge variety of plants that can support a large range of animal life. These animals range from tigers to brightly colored birds, such as the toucan (right).

DESERTS

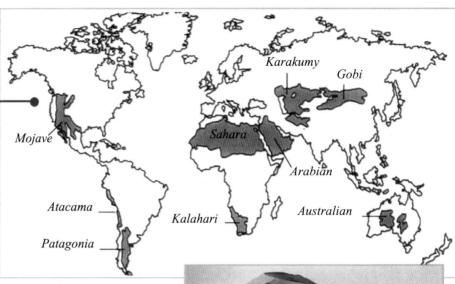

DESERTS

Although generally thought of as hot areas, the word "desert" applies to any region with very little rainfall. Deserts cover about 14 percent of the Earth's land area, the largest being the Sahara in North Africa. Of these deserts, only 10 to 20 percent are actually covered in sand.

Sand dunes
Dunes (right) are formed by wind blowing over sand in the same way that wind causes waves when it blows over water. Sand dunes can be up to 1,525 feet (465 m) high and 3 miles (5 km) long.

Animals of the desert
Creatures that have to live in the desert have developed special physical features and actions that suit the harsh climate. Camels (top) have large, padded feet to stop them from sinking into the sand and long eyelashes to stop sand from being blown into their eyes. The Fennec fox (right) has large ears to help it keep cool. Desert lizards (above) need the warm weather to survive. However, too much heat can be fatal. To escape the harsh sun, the lizard burrows into the cool ground.

RAIN SHADOW

Some deserts, such as the Atacama in South America, are called rain-shadow deserts. They are usually found behind a chain of mountains. As warm, moist air blows into the mountain chain, it rises and cools. As it cools, the air is unable to hold onto its water, which forms clouds and then falls as rain. As the now-dry air passes over the mountains, it falls and gets warmer, creating a warm, dry area, or rain-shadow zone (above).

ICE SHEETS *and* TUNDRA

THE POLES
The extremes at either end of the Earth are very cold places that have to endure extended periods of darkness during the winter. The landscape here is dominated by large, moving ice sheets. Farther away from the Poles, the ice sheets give way to tundra, a treeless landscape, where only grasses and mosses can grow above a permanently frozen subsoil.

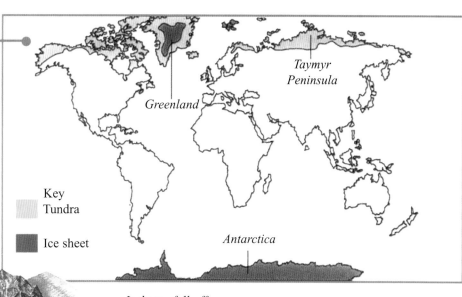

Key
Tundra
Ice sheet

Greenland
Taymyr Peninsula
Antarctica

Glacier
Icebergs fall off
Iceberg
Cracks multiply as ice nears sea

Glaciers
Glaciers spread out from a source. As the glacier approaches the warmer waters of the sea, cracks form. These cracks grow and eventually large chunks of ice fall off to form icebergs (left).

Animals of the cold
Many animals, such as the Arctic fox (above), change the color of their coats with the seasons. In summer they are dark, while in winter they are white to match the snow. The musk ox (right) has a large, shaggy coat to keep it warm.

ICE SHEETS
The area around the North Pole, the Arctic, has no land. Instead, the Arctic Ocean is covered by an immense ice sheet, which varies in size with the changing seasons. The South Pole, or Antarctica, is the coldest continent on the planet. Like the Arctic, the area is covered by thick sheets of ice all year round. However, beneath the ice is a large landmass, complete with mountain ranges (see page 37).

TUNDRA
Because the tundra regions have long, cold winters and short summers, trees cannot grow in these regions (left). About 12 inches (30 cm) below the surface, the soil is frozen all year round. This frozen soil is known as "permafrost" and it stops water draining from the surface, keeping the soil boggy. Other tundra features include pingoes.

Ice
Gas pressure

Pingoes
Pingoes are bulges in the ground caused by the earth being forced up by a buildup of gas beneath the ice.

GRASSLANDS

The world's grasslands
The steppes of Asia cover an enormous area, stretching from the Ukraine to Siberia. The world's largest prairie region is in North America, while almost two-fifths of Africa is covered by savanna (left).

GRASSLANDS

Grasslands occupy large areas of land within the larger continents. Examples include the steppes of Central Asia, the prairies of North America, the Pampas of South America, and the savanna of southern Africa. There are three types of grassland (steppe, prairie, and savanna—see below) and they often lie between desert and rich forested areas. Much of these grasslands has been turned over to farming, usually grazing livestock.

WILDLIFE

Despite the lack of significant tree cover, grasslands hold a rich and diverse range of animal life. These include herds of grazing animals that migrate across them in search of food, such as the giraffe (above) and zebra (below) of Africa and the pronghorn of North America. Animals that do not migrate also live in grasslands. Prairie dogs (bottom left) live in burrows on the North American Prairie.

TYPES OF GRASSLAND

Savanna is scattered with trees and shrubs, including acacias and baobabs. Its grass does not totally cover the ground, growing, instead, in clumps. These regions may also endure long drought periods. Prairies have often been referred to as "seas of grass." They are covered in long grasses with flowers, while trees such as cottonwood and willows are restricted to river valleys. The steppes are covered in very short grasses and found in dry areas with hot summers and cold winters.

Area shown by map

The MAPS

The following section looks at the world in more detail. Each chapter examines a different region, from North America to Australia and the Poles. Comprehensive maps reveal each region's physical features, and highlight a few cultural, economic, and political points of interest. Each map also shows the flags of the nations, their capitals, and other major cities. Accompanying text describes the region's countryside and points out individual physical features, such as the length of major rivers and the altitude of the highest mountains, to present a full view of the world's continents.

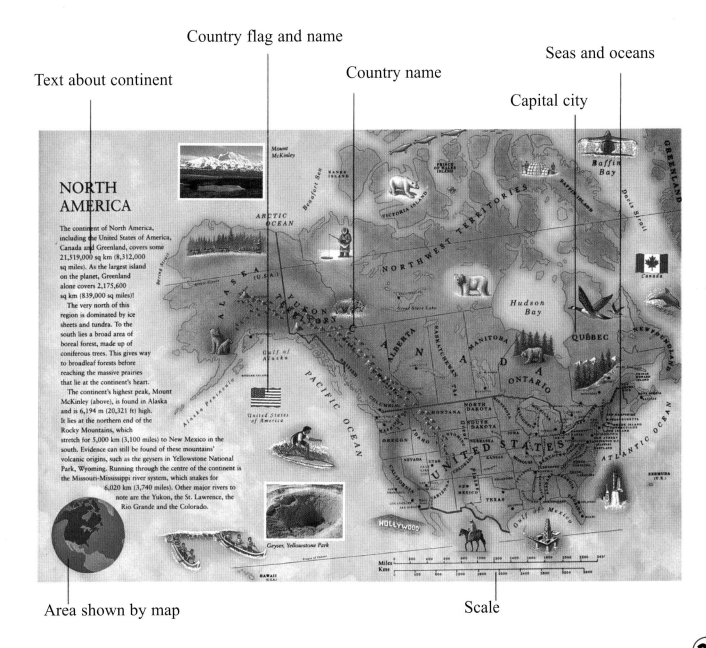

Text about continent

Country flag and name

Country name

Seas and oceans

Capital city

NORTH AMERICA

The continent of North America, including the United States of America, Canada and Greenland, covers some 21,519,000 sq km (8,312,000 sq miles). As the largest island on the planet, Greenland alone covers 2,175,600 sq km (839,000 sq miles)!

The very north of this region is dominated by ice sheets and tundra. To the south lies a broad area of boreal forest, made up of coniferous trees. This gives way to broadleaf forests before reaching the massive prairies that lie at the continent's heart.

The continent's highest peak, Mount McKinley (above), is found in Alaska and is 6,194 m (20,321 ft) high. It lies at the northern end of the Rocky Mountains, which stretch for 5,000 km (3,100 miles) to New Mexico in the south. Evidence can still be found of these mountains' volcanic origins, such as the geysers in Yellowstone National Park, Wyoming. Running through the centre of the continent is the Missouri-Mississippi river system, which snakes for 6,020 km (3,740 miles). Other major rivers to note are the Yukon, the St. Lawrence, the Rio Grande and the Colorado.

Geyser, Yellowstone Park

Area shown by map

Scale

NORTH AMERICA

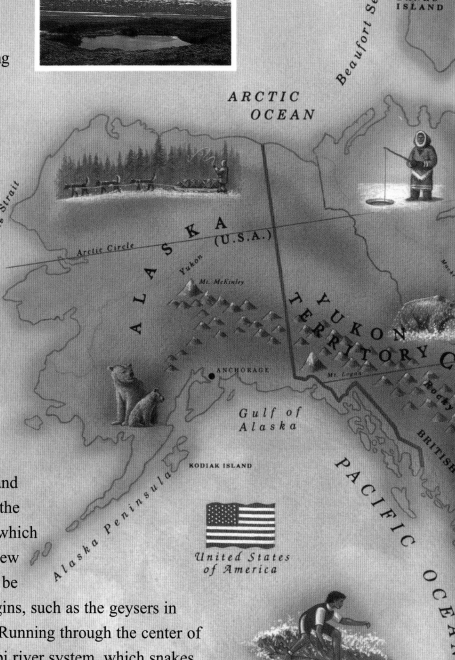

Mount McKinley

The continent of North America, including the United States, Canada, Greenland, Mexico, Central America, and the islands of the Caribbean, covers about 15,060,000 square miles (39,000,000 sq km). The largest island on the planet, Greenland, covers 840,000 square miles (2,175,600 sq km)!

The very north of this region is dominated by ice sheets and tundra. To the south lies a broad area of boreal forest, made up of coniferous trees. This gives way to broadleaf forests before reaching the massive prairies that lie at the continent's heart.

The continent's highest peak, Mount McKinley (above), is found in Alaska and is 20,321 feet (6,194 m) high. It lies at the northern end of the Rocky Mountains, which stretch for 3,125 miles (5,000 km) to New Mexico in the south. Evidence can still be found of these mountains' volcanic origins, such as the geysers in Yellowstone National Park, Wyoming. Running through the center of the continent is the Missouri-Mississippi river system, which snakes for 3,740 miles (6,020 km). Other major rivers to note are the Yukon, the St. Lawrence, the Rio Grande, and the Colorado.

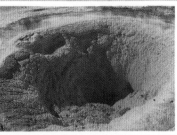

Geyser, Yellowstone National Park

SOUTH and CENTRAL AMERICA

Linking the two continents of North and South America is the narrow strip of land called Central America. This region's countryside changes from scrub and desert in the north of Mexico to lush, tropical rainforest, which stretches from the Yucatan Peninsula all the way to South America.

To the east is the Caribbean Sea, which is home to hundreds of islands. These lie in an arc from Cuba and the Bahamas in the north, to Trinidad and Tobago just off the coast of Venezuela.

South America itself contains one of the world's largest river systems. The Amazon winds for 4,073 miles (6,555 km) from the Andes to the Atlantic. Its river basin covers 4.5 million square miles (11.6 million sq km), has more than 200 tributaries, and is mostly covered in dense rainforest (see page 17). South America also has the world's longest mountain chain, the Andes, stretching for 4,500 miles (7,242 km) from Panama to Cape Horn. They include the continent's tallest peak, Aconcagua, whose summit is 22,834 feet (6,960 m) high.

Ecuador

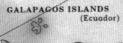
Peru

Bolivia

Chile

Argentina

The Andes

Paraguay

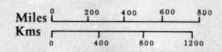

EUROPE

European farmland

Europe, including the Russian Federation to the Urals, covers 4,052,000 square miles (10,495,000 sq km).

The far north of the continent lies within the Arctic Circle, which is dominated by tundra (see page 19). Farther south, the majority of the continent is taken up by the European plain, a large belt of fertile land, much of which has been turned over to farming (above).

To the south, the region is covered by Mediterranean scrubland.

Major European mountain chains include the Urals, which divide the continent from Asia, the Pyrenees between France and Spain, the Alps in Central Europe, and the Carpathians to the east. Major rivers include the Rhine, the Rhône, the Danube, the Volga, and the Vistula.

AFRICA and the MIDDLE EAST

African savanna

Africa covers 22 percent of the Earth's land area—about 11,710,500 square miles (30,330,000 sq km). The northern half of the continent is dominated by the Sahara Desert, which stretches from the Atlantic Coast to the Red Sea. To the south is the area of grassland called the "Sahel." The continent's center contains the Congo river basin and the rainforest that fills the area. To the east of this are the savanna of the Masai Mara (above) and the Great Rift Valley. This is a series of cracks in the earth that has formed steep-sided valleys which run from Mozambique to Syria.

The continent's major rivers include the Niger, the Congo, and the Nile, which stretches for 4,160 miles (6,695 km). Major mountain chains include the Atlas Mountains and the Ethiopian Highlands. The highest point is Mount Kilimanjaro in Tanzania, which is 19,340 feet (5,895 m) high.

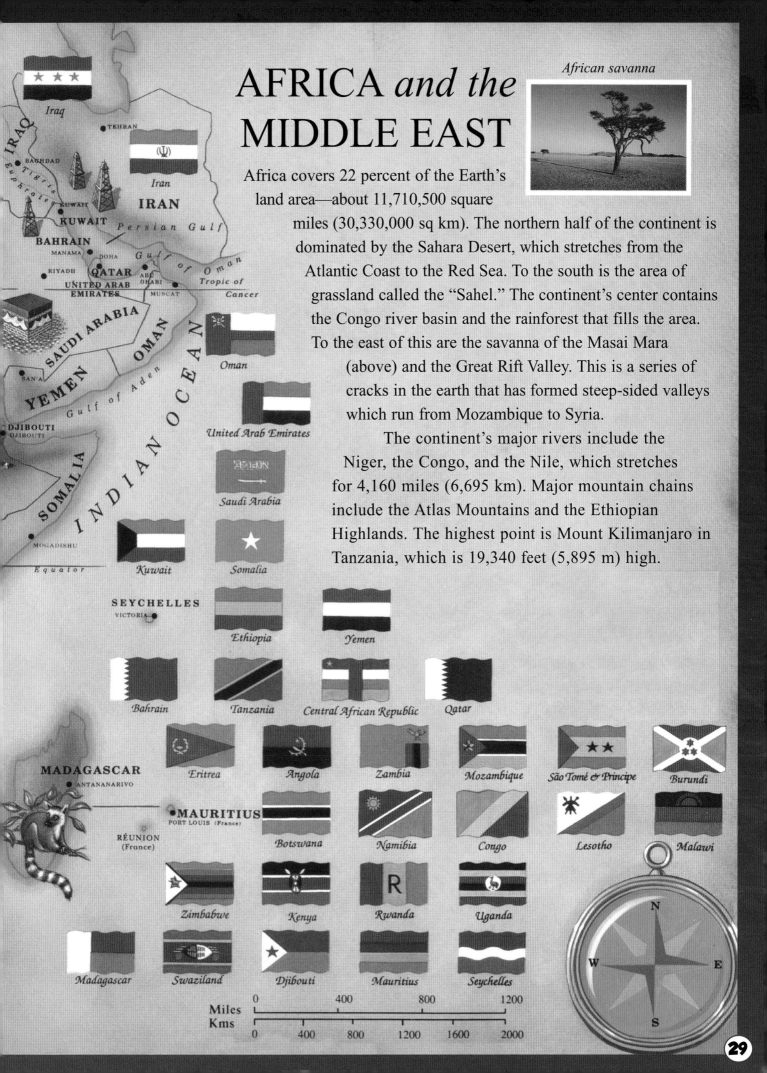

Mount Fuji, Japan

NORTHERN ASIA

Sitting within the Arctic Circle, the north of this continent is covered in tundra (see page 19). Below this lies a large band of boreal forest, or taiga (see page 17). Sitting in the center of the continent is the Gobi Desert, which covers 50,781 square miles (1.3 million sq km). Surrounding this dry area are wide tracts of grasslands, known as steppes (see page 20).

Major rivers include the Volga and the Chang Jiang (Yangtze), which is 3,430 miles (5,520 km) long.

The eastern coast of Asia sits on the edge of the Ring of Fire.

It has seen a great deal of volcanic activity, that has over millions of years created peaks such as Mount Fuji in Japan (above).

SOUTH and SOUTHEAST ASIA

South and Southeast Asia are lush and fertile places. Apart from the desert area in western Pakistan and the scrublands of Central India, the region is rich in fertile farmland (left) and tropical vegetation.

The major rivers include the Ganges, which flows from the Himalayas to the Indian Ocean at Bangladesh, and the Mekong, which is 2,749 miles (4,425 km) long.

The major mountain chain in this region is the range called the Himalayas. These lie just south of the Himalayan Plateau and stretch for over 1,500 miles (2,400 km). They contain the world's highest peak—Mt. Everest, which is 29,028 feet (8,848 m) high. The region of sea between the Pacific and the Indian Ocean is filled with thousands of islands—Indonesia alone is made up of over 13,600 islands, including Java and Sumatra.

Terraced rice paddies

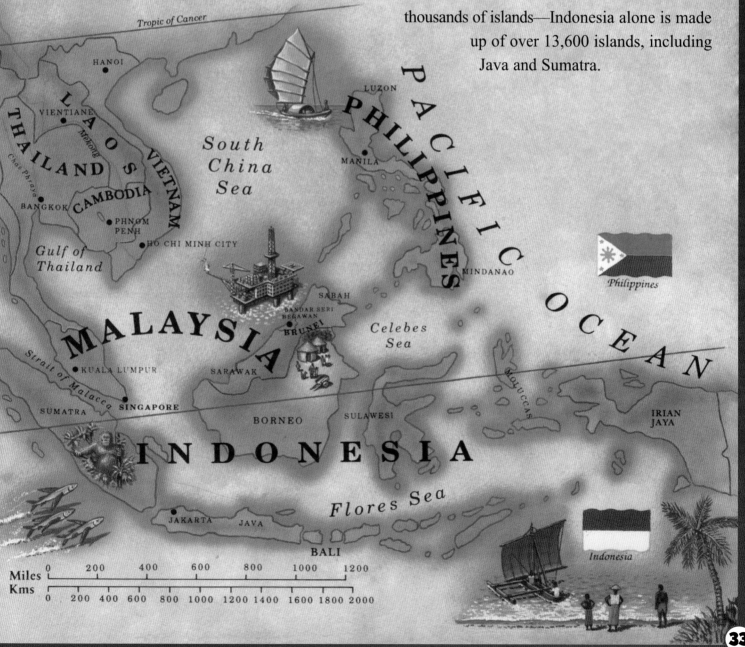

The PACIFIC ISLANDS and AUSTRALIA

Coral atoll

The country of Australia forms the world's smallest continent, covering 2,966,200 square miles (7,682,300 sq km). The majority of the country's interior is covered in scrub and desert, notably the Great Sandy and the Great Victoria Deserts. To the north, the climate is more tropical, supporting lush rainforest. Running north to south along the eastern edge of the country is the Great Dividing Range. At the southern end of this range is the continent's highest point, Mount Kosciusko, which is 7,316 feet (2,230 m). The longest river is the Murray-Darling system, at 2,310 miles (3,717 km) long. New Zealand's climate is far cooler than Australia's—its southern coast reflects the fjord landscape of northern Europe. The Pacific Ocean holds hundreds of tiny islands, such as those of Western Samoa. Many of these islands are coral atolls—huge structures formed by the skeletons of tiny animals.

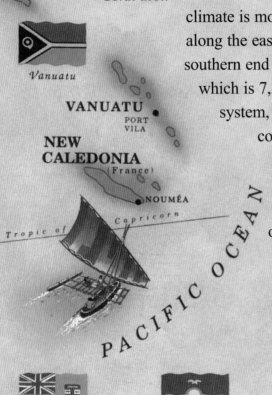

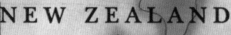

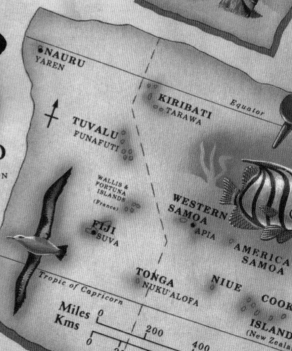

ANTARCTICA

Antarctic Peninsula

This continent, including the ice cap that stretches into the Southern Ocean in many places, covers about 5,468,750 square miles (14,000,000 sq km). The ice sheet averages about 6,800 feet (2,100 m) thick, and can reach 14,800 feet (4,500 m) deep.

The continent is also the place where the world's lowest natural temperature was recorded—an astonishing -128.6°F (-89.2°C). Normally, the temperature rarely reaches above 32°F (0°C). Antarctica has no rivers, but does contain mountain chains such as the Transantarctic Mountains and the highest point, the Vinson Massif, at 15,900 feet (4,840 m) high.

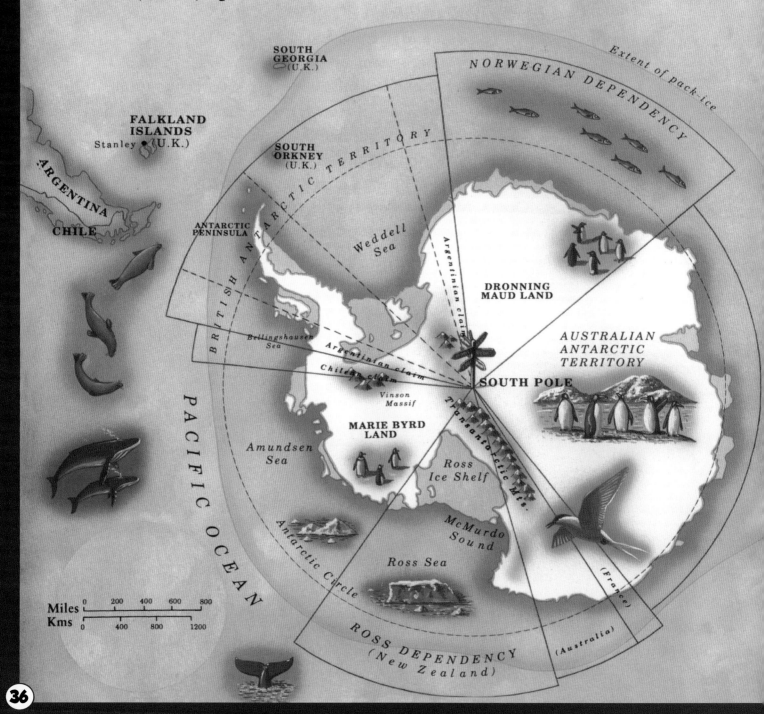

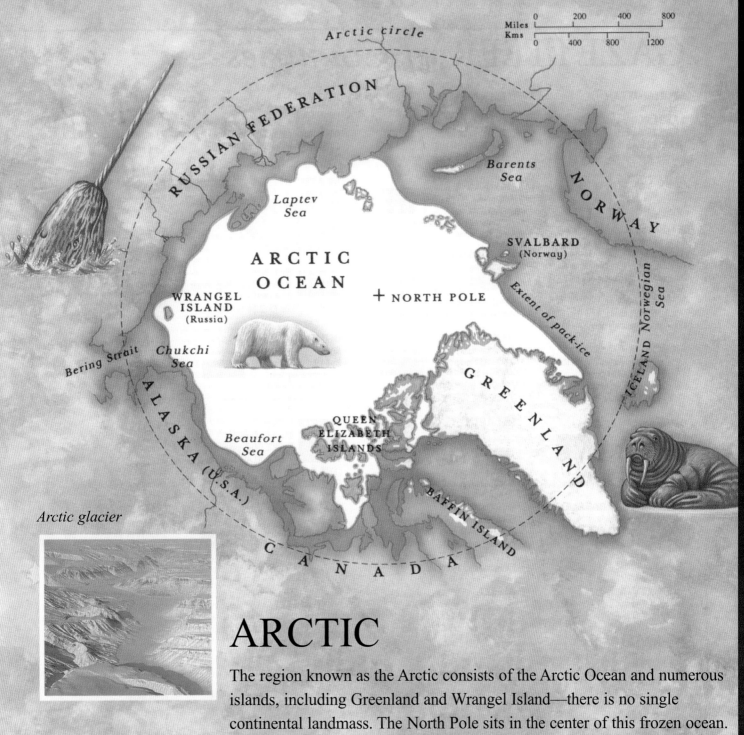

Arctic glacier

ARCTIC

The region known as the Arctic consists of the Arctic Ocean and numerous islands, including Greenland and Wrangel Island—there is no single continental landmass. The North Pole sits in the center of this frozen ocean. The Arctic Ocean, the world's smallest ocean, covers about 5,440,000 square miles (14,090,000 sq km), and most of this lies under ice throughout the year. This pack ice increases in size during the winter months, when most of the region is in darkness while the sun remains below the horizon. This is because the Earth spins around the sun at an angle, keeping the North Pole in shade during the winter months. Beyond the ice sheet, the surrounding islands are covered with a tundra landscape (see page 19), which frequently blooms with flowers during the milder summer months.

I KNOW ABOUT!
WORLD OF WONDER

SHARKS

44 WHAT IS A SHARK?

WHERE DO SHARKS LIVE? 46

IT'S A SHARKS LIFE! 48

DANGEROUS SHARKS
54

STRANGE SHARKS
58

SHARK RELATIVES
60

MORE FASCINATING FACTS
64

All kinds of sharks

Sharks are meat-eating fish that are found all around the world. There are more than 350 different kinds of sharks. They range in size from the huge whale shark to tiny sharks that can fit in the palm of your hand. Although many sharks are fierce hunters, the biggest sharks, like the whale shark and the basking shark, are gentle creatures. Basking sharks eat only the tiny plants and animals that drift in the sea.

Tiny sharks
The tiny dwarf shark is only about 5.5 inches (14 cm) long when fully grown. Lantern sharks and dogfish are other kinds of small sharks.

LEOPARD SHARK

Huge sharks
Whale sharks are the biggest fish in the world. An adult whale shark can be up to 40 feet (12 m) long. They may weigh more than 14 tons (12 t), which is about twice as heavy as an adult African elephant.

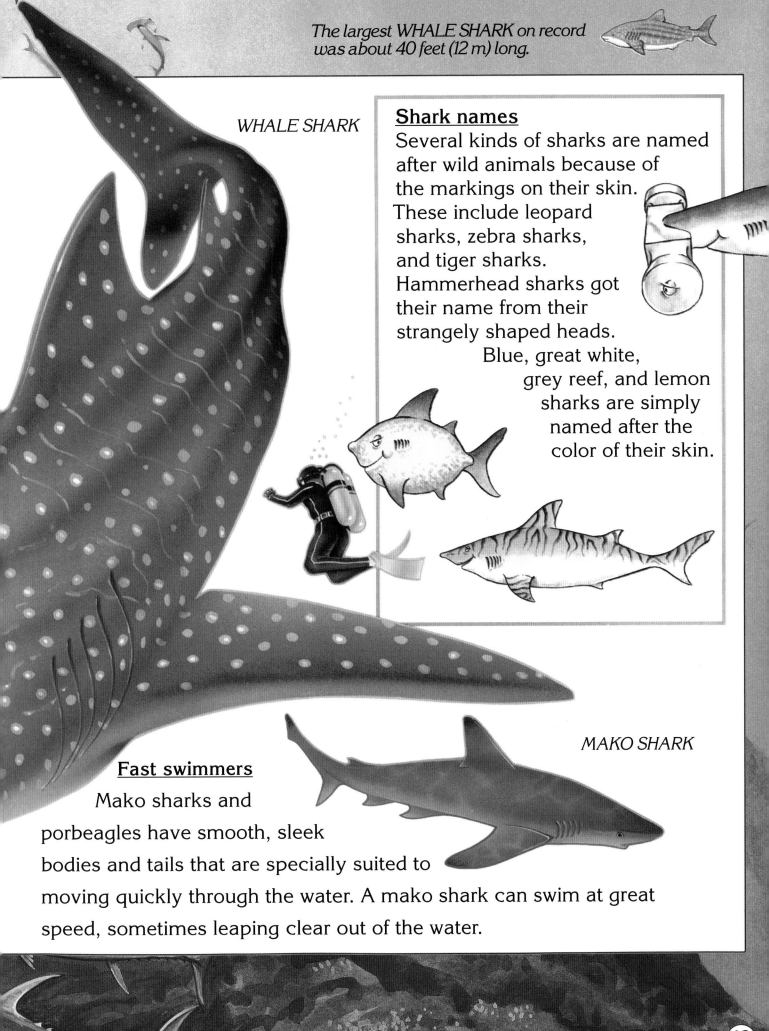

The largest WHALE SHARK on record was about 40 feet (12 m) long.

WHALE SHARK

Shark names
Several kinds of sharks are named after wild animals because of the markings on their skin. These include leopard sharks, zebra sharks, and tiger sharks. Hammerhead sharks got their name from their strangely shaped heads. Blue, great white, grey reef, and lemon sharks are simply named after the color of their skin.

MAKO SHARK

Fast swimmers
Mako sharks and porbeagles have smooth, sleek bodies and tails that are specially suited to moving quickly through the water. A mako shark can swim at great speed, sometimes leaping clear out of the water.

What is a shark?

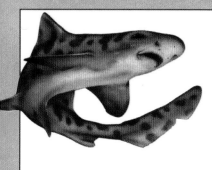

Sharks are some of the most powerful and fearsome creatures in the sea. They come in lots of different shapes and sizes. Many have streamlined bodies that help them swim quickly and easily. Sharks do not have scales like bony fish. Instead, their bodies are rough and covered in tiny, thornlike points called "denticles," or skin-teeth.

Skeletons
Most fish have a skeleton made of bones. But a shark's skeleton is made of a tough, flexible material called "cartilage." Bony fish also have a special gas-filled bag, called a "swim bladder," to keep them afloat in the water. Sharks don't have a swim bladder, and have to keep swimming to avoid sinking.

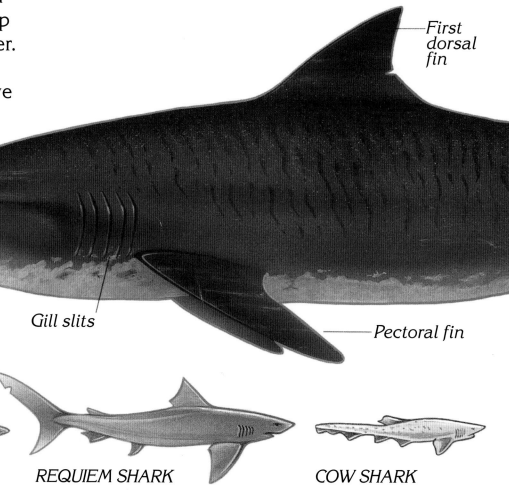

TIGER SHARK

First dorsal fin

Gill slits

Pectoral fin

BRAMBLE SHARK

REQUIEM SHARK

COW SHARK

MAKO SHARKS can swim at speeds of more than 35 miles (56 km) per hour.

Fins

Most sharks have two sets of paired fins, as well as two fins on their back and a smaller fin on the underside of the body. The large triangular-shaped fin (dorsal fin) on a shark's back can sometimes be seen above the water's surface. Fins help the shark to balance and steer as it swims through the water.

Streamlining

The streamlined shape of most sharks helps them to swim easily through the water. A streamlined shape is rounded and blunt at the front, and pointed at the back. Water flows smoothly past it. Use stiff cardboard to make a shape like the one below, as well as a rounded shape like a cylinder. Test to see which shape moves more easily through water.

Denticles

Second dorsal fin

Caudal fin

Pelvic fin

Anal fin

Shark tails

Sharks bend their tails from side to side to push their bodies through the water.

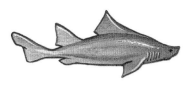

ROUGH SHARK   THRESHER SHARK   BLIND SHARK

WHERE DO SHARKS LIVE?

Sharks live in the world's oceans and seas, from the chilly Arctic Ocean to the warm tropical waters off Africa. Some sharks live in the deepest part of the ocean, while others spend most of their time near the surface. Nurse sharks stay near the coast, but most sharks live far out to sea.

The greatest traveler
Blue sharks travel long distances in the tropical waters and warm, temperate seas of the Atlantic Ocean. Blue sharks from the north Atlantic have been found more than 3,700 miles (5,954 km) away, off the coast of South America.

Above and below
Basking sharks and thresher sharks swim along the water's surface looking for food. Carpet sharks and horn sharks live on the seabed. Swell sharks rest on the ocean floor during the day, and hunt for food at night. Hammerhead sharks often come to the surface in groups of 100 or more, before swimming off to another part of the ocean.

SLEEPER SHARKS in the Pacific Ocean can live more than 6,500 feet (1,981 m) below the surface.

Island legends

People on many of the islands in the Pacific Ocean have worshipped sharks for thousands of years. In New Guinea, no one was allowed to kill a shark, in order to avoid offending the gods of the sea. The people of Hawaii worshipped a shark king called "Kamo Hoa Lii." Some Pacific Islanders believed that sharks contained the spirits of their dead relatives.

SEA SPIRIT

Leaving the sea

Bull sharks sometimes leave the sea altogether and swim into freshwater rivers and lakes. They have been found in lakes and rivers in Central and South America, India, and the United States. Bull sharks have been spotted in the River Zambezi in Africa, at least 125 miles (201 km) away from the sea.

It's a shark's life!

Sharks can see, hear, smell, taste, and touch—just like people. Hammerhead sharks have good eyesight and can see clearly in the deep, dark waters of the ocean. They can see all around when swimming along because their eyes are set on the tips of their hammer-shaped heads. Sharks also have a very good sense of hearing. They can only hear low-pitched noises, but these sounds travel long distances underwater.

Feeling for food

Barbels

Nurse sharks, wobbegongs, and bamboo sharks (above) all have special feelers, called "barbels," on either side of their nose. They use them to feel their way along the seabed in search of food.

Special senses

A special sensing line runs along both sides of a shark's body, from its head all the way down to its tail. This is called a "lateral line." It helps the shark to detect prey, and anything moving in the water, such as other fish and sharks.

Nostril

Pores

Tiny skin holes, or pores, around a shark's head pick up electric signals from other fish, helping the shark to find its prey.

Sharks have a THIRD EYELID that protects their eyes when they bite.

LATERAL LINE

A GREAT WHITE SHARK can sense one drop of blood in 25 gallons (100 liters) of water.

Staying alive

Like all animals, sharks need oxygen to stay alive. They get their oxygen from sea water. The water goes in through their mouth and flows over tiny featherlike parts behind the skull, called "gills." The gills remove the oxygen. When sharks swim, water flows over their gills and out through their gill slits. Most sharks have five pairs of gills. Angel sharks, which live on the seabed, get their oxygen by taking in water through special holes on the top of their head.

Shark friends

Fish called "remoras" attach themselves to sharks with a special sucker on the top of their head. They get a free ride from sharks, but in return they eat the tiny creatures living in the sharks' skin.

Small, stripy, pilot fish also swim close to sharks. People used to think that these fish guided the sharks toward their food. In fact, they probably swim along in shoals beside sharks for protection.

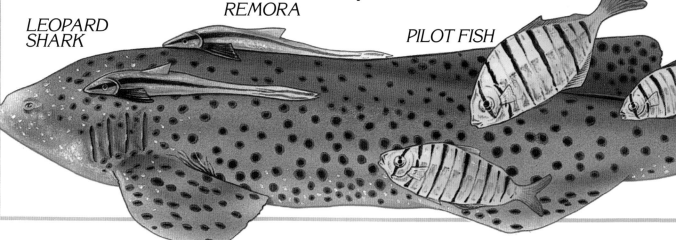

LEOPARD SHARK

REMORA

PILOT FISH

FEEDING

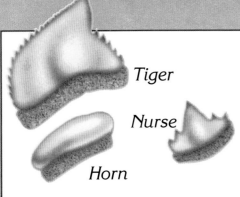

Tiger
Nurse
Horn

Teeth
Sharks have different kinds of teeth to suit the food they eat. Horn sharks have strong, flat teeth to crush the shells of crabs and other shellfish. Tiger sharks have sharp, jagged teeth to stab at and tear off pieces of food.

Most sharks use their sharp teeth to eat fish and other smaller kinds of shark. But the biggest sharks of all, the basking sharks and whale sharks, have a very different diet from most sharks. They feed on tiny animals and plants, called "plankton," that float in the sea.

Most sharks have several rows of teeth. Broken teeth are quickly replaced by new, sharp ones.

Power jaws
When a shark bites, its jaw moves forward so that its teeth stick out. This makes it easier for the shark to snatch its prey. The shark then grips its victim in between its powerful jaws, to stop it from escaping. Although the jaws of the megamouth shark measure more than one yard (almost one meter) across, this gentle giant eats tiny shrimps and plankton.

A shark may have up to 3,000 TEETH inside its mouth at one time.

Filter feeding

Basking sharks and whale sharks take in huge mouthfuls of water and filter out the plankton. Because of the way they feed, these sharks are called "filter feeders."

BASKING SHARK

What's on the menu? Shark food ranges from tiny plankton to mammals, such as porpoises and sea-lions. As well as eating live fish, larger sharks eat sea creatures like dolphins, seals, and turtles, as well as seabirds.

SEA URCHIN

Strange things to eat

The tiger shark is sometimes called "the trashcan of the sea." As well as eating poisonous sea snakes, hard-shelled sea turtles, or jellyfish, they also eat trash that has been thrown overboard from ships or has floated out to sea from the land.

Old tin cans, dogs, pieces of coal, and cardboard cartons have all been found inside tiger sharks.

TURTLE

In a FEEDING FRENZY, sharks get overexcited by the smell of blood in the water and may attack each other.

JELLYFISH

Eggs and babies

Most sharks give birth to live babies called "pups." The eggs hatch inside the mother's body, and the newborn sharks come out alive. Other sharks, like dogfish, lay eggs. Once the mother shark has laid her eggs, she swims away, leaving the baby sharks to hatch on their own.

Finding a mate
Some sharks travel to special breeding grounds to find a mate. A male white-tip reef shark (above) chases after a female, playfully biting or grabbing hold of her fins with his teeth.

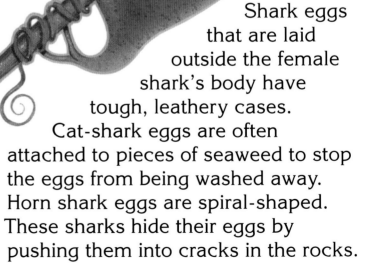

A baby SWELL SHARK is developing inside this leathery egg case.

Eggs
Shark eggs that are laid outside the female shark's body have tough, leathery cases. Cat-shark eggs are often attached to pieces of seaweed to stop the eggs from being washed away. Horn shark eggs are spiral-shaped. These sharks hide their eggs by pushing them into cracks in the rocks.

The eggs of the WHALE SHARK can be up to 11 inches (28 cm) long.

Mermaid's purses

Mermaids have appeared in stories about the sea for hundreds of years. The top half of a mermaid is like a woman, and the bottom half is like a fish. The empty egg cases of sharks that are washed up on the seashore are often called "mermaid's purses" because of their strange color and shape.

Egg case of a CAT SHARK

HORN SHARK egg case

Live young

Some sharks give birth to just one or two pups. Others, like hammerhead sharks, may have up to 40 pups at a time. When lemon sharks are born, they are attached to the mother shark by a special cord. Human babies are attached to their mothers in the same way.

LEMON SHARK

Dangerous sharks

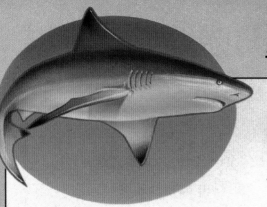

Only about 50 of the 350 kinds of sharks are known to attack people in the water. Sharks mainly harm swimmers, surfers, and deep-sea divers. Some scientists think that sharks may mistake these people for seals or large fish. The most dangerous sharks are the great whites. These fierce and powerful hunters can be more than 20 feet (6 m) long, and they have large, razor-sharp teeth.

Keeping safe
Some sharks swim close to the shore and attack people in the shallow water. In vacation resorts in Australia, South Africa, and along the west coast of the United States, bathers are protected from sharks by nets. These nets, which are made from steel mesh or from chains, help to prevent the sharks from swimming near to the shore.

In Sri Lanka, SNAKE CHARMERS were used to keep local sharks away from the pearl divers.

Danger!

Other dangerous sharks include bull sharks, tiger sharks, and grey reef sharks. When angry, the grey reef shark arches its back and lowers the fins on the sides of its body. It behaves in this way when it is just about to attack its victim.

About 100 SHARK ATTACKS take place around the world each year. Many people survive these attacks, but they are sometimes left with terrible injuries.

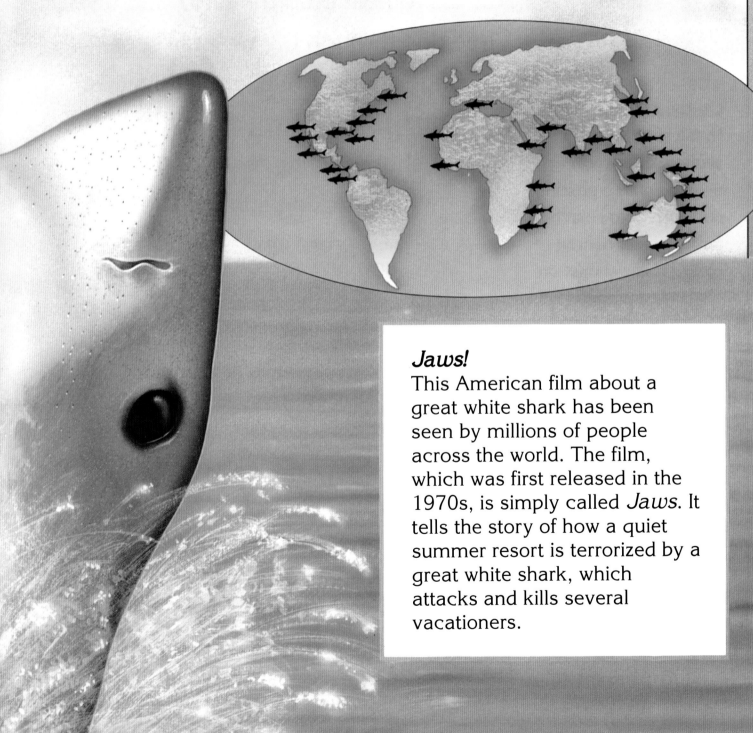

Jaws!

This American film about a great white shark has been seen by millions of people across the world. The film, which was first released in the 1970s, is simply called *Jaws*. It tells the story of how a quiet summer resort is terrorized by a great white shark, which attacks and kills several vacationers.

HARMLESS SHARKS

Most sharks are not dangerous at all. In fact, the biggest sharks of all, the whale sharks and basking sharks, are quite harmless. They will even allow divers to hold onto their fins and take a free ride through the water. These huge sharks are filter feeders (see page 51) and they do not have the sharp cutting teeth of the fierce attacking sharks, like the great white.

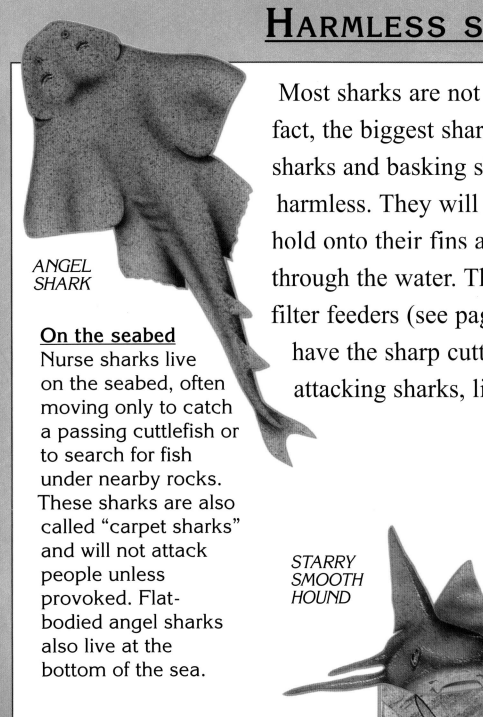

ANGEL SHARK

On the seabed
Nurse sharks live on the seabed, often moving only to catch a passing cuttlefish or to search for fish under nearby rocks. These sharks are also called "carpet sharks" and will not attack people unless provoked. Flat-bodied angel sharks also live at the bottom of the sea.

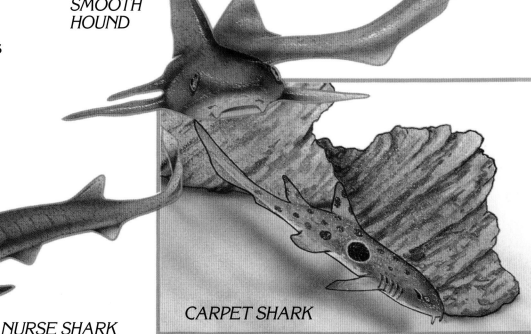

STARRY SMOOTH HOUND

NURSE SHARK

CARPET SHARK

The teeth of the BASKING SHARK are the same size as grape seeds.

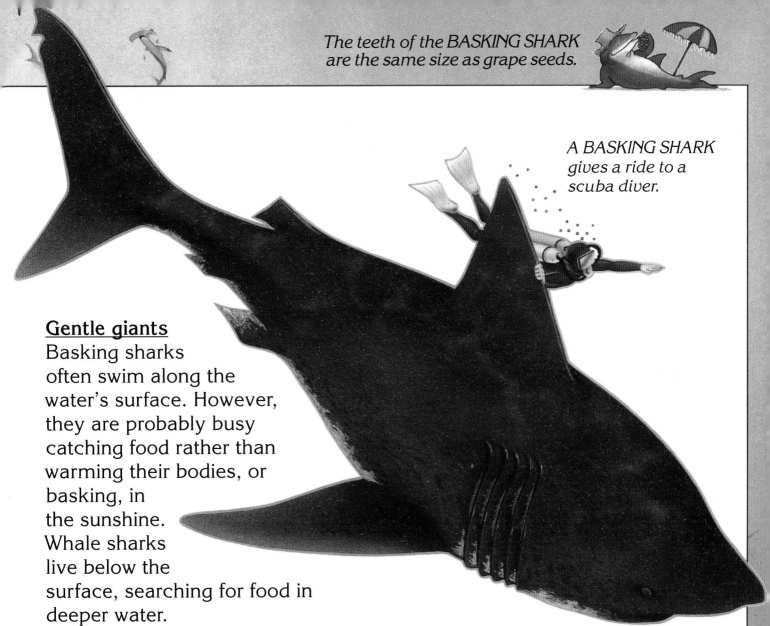

A BASKING SHARK gives a ride to a scuba diver.

Gentle giants
Basking sharks often swim along the water's surface. However, they are probably busy catching food rather than warming their bodies, or basking, in the sunshine. Whale sharks live below the surface, searching for food in deeper water.

What is camouflage?
Camouflage means being able to hide easily from others by blending in with your surroundings. On the seabed, the speckled skin of the angel shark helps it to hide in the sand. Swell sharks gulp down mouthfuls of water or air and then squeeze their bodies between the rocks.

SWELL SHARK

Strange sharks

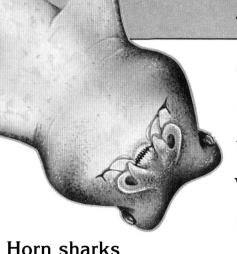

One of the strangest-looking groups of sharks are the hammerheads. These odd fish have their eyes and nostrils at the ends of their wide, hammerlike heads. Another unusual shark is the frill shark, which has a collar of skin folds behind its head. This shark is also known as the "lizard-head."

Horn sharks
The Port Jackson shark (above) is one kind of horn shark. It lives in the warm waters of the Indian and Pacific Oceans. Port Jackson sharks are also known as "pig sharks" because of their square-shaped head and large nostrils.

HAMMERHEAD SHARK

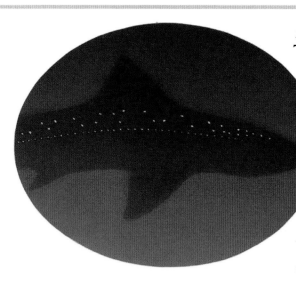

Shining in the dark
The tiny lantern shark lives deep down in the darkest parts of the ocean. Its small body glows in the dark water, even as far down as 6,500 feet (1,981 m) below the surface. Lantern sharks have special parts on their bellies that give off light. Other sharks that glow in the dark are cookie cutters and green dogfish.

SWELL SHARKS can blow up to double their usual size when they are frightened.

Rare sharks

Scientists know very little about some sharks, because they have only ever seen a small number of them. One of the rarest sharks is the goblin, or elfin shark, which has a long, pointed horn on the top of its head. A completely new kind of shark, the megamouth, was found off the coast of Hawaii in 1976. Fewer than 10 of these enormous sharks have ever been seen.

This rare GOBLIN SHARK lives in very deep water.

In disguise

Most carpet or nurse sharks live on the ocean floor in tropical seas. One of these is the wobbegong, which lives off the coasts of Australia as well as several Asian countries, including China, Vietnam, and Japan. Wobbegong is the shark's Aborigine name. These sharks are very well camouflaged. The blotchy markings and the color of their skin blend in with the rocks and corals around them. Even the bearded fringe below the wobbegong's mouth looks like strands of seaweed.

The ornate WOBBEGONG can easily hide among the corals and seaweed on the seabed.

Shark relatives

SHARK FOSSIL

The first sharks lived about 400 million years ago. But most of today's sharks are more like the ones that roamed the seas after the dinosaurs disappeared, about 65 million years ago. The giant megalodon shark, which was about 43 feet (13 m) long, ruled the seas about 20 million years ago. Megalodons may still have been alive 12,000 years ago. Its closest relative alive today is the great white shark.

Ancient sharks
The remains of the earliest known sharks have been found in the United States. When the sharks died, their bodies sank down onto the seabed and were preserved in layers of sand, mud, and rock. The buried remains of once-living things are known as "fossils."

Flying rays
Rays use the huge wing-like fins on the sides of their body to swim through the water. They look as if they are really "flying" through the sea. They are filter feeders like basking sharks, eating the tiny plankton in the sea.

One of the most common of the early sharks, called CLADOSELACHE, lived around 350 million years ago.

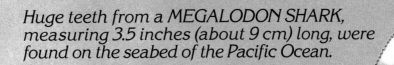
Huge teeth from a MEGALODON SHARK, measuring 3.5 inches (about 9 cm) long, were found on the seabed of the Pacific Ocean.

Shark cousins

Rays, skates, and sawfish belong to the same group of fish as sharks. Rays spend most of their time on the seabed, while sawfish are often found in freshwater rivers and lakes. The body of the ray is so broad that the width of its body is more than the length. Giant manta rays are sometimes more than 20 feet (6 m) wide.

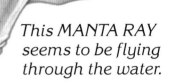
This MANTA RAY seems to be flying through the water.

The sting of the stingray

Stingrays were given their name because of the poisonous spines in their tails. These spines have tiny barbs along the edges. Stingrays use these spines to defend themselves against enemies, such as hammerhead sharks.

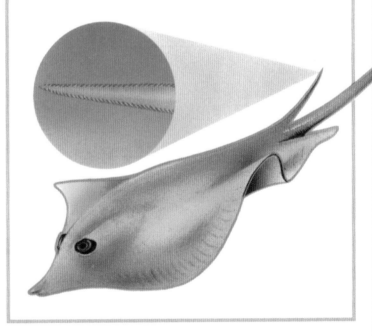

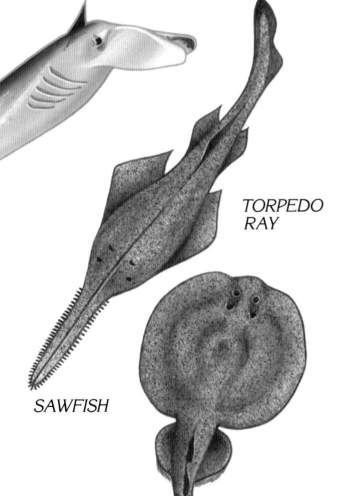

TORPEDO RAY

SAWFISH

SHARKS AND PEOPLE

Food
The Chinese make a special soup from dried shark fins. Shark meat is eaten by people in many places around the world. In England, fried spiny dogfish (above) is sold in fish and chip shops.

All over the world, people have found different ways to make use of sharks. We capture sharks and take their flesh, skin, teeth, fins, and even their liver. Sharks provide people with food to eat, with skin to make into leather goods, with teeth to make into jewelry and decorations, and with oil that is used to manufacture skin creams, lipsticks, and pills. Shark fishing is a popular sport in the Caribbean and off the coasts of Australia and North America.

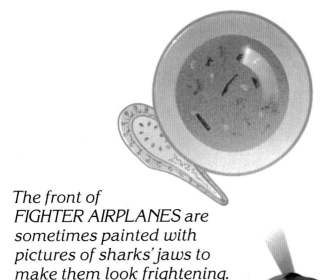

SHARK FIN soup

The front of FIGHTER AIRPLANES are sometimes painted with pictures of sharks' jaws to make them look frightening.

Sharks and medicine
In the past, the Romans treated teething pains by rubbing their children's teeth with dried shark brains. Scientists have found that sharks very rarely develop cancer. Today, cancer kills more than 5 million people in the world each year. Research into sharks may help doctors find a cure for this killer disease.

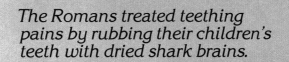

The Romans treated teething pains by rubbing their children's teeth with dried shark brains.

Sharks' enemies

Human beings are sharks' main enemies. Sharks are killed for sport or for their meat and oil. There is a danger that people are killing too many sharks, and some kinds may even disappear altogether. Dolphins and porpoises also attack sharks, and occasionally win the fight.

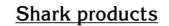

Shark products

Shark skin is made into a whole range of products, from boots and belts to sword handles and leather boxes. It is dyed and then polished to make it smooth. Rough shark skin is called "shagreen." It used to be used to smooth down wood and precious stones. Many Japanese people take shark oil pills to prevent heart disease and cancer. The oil, which comes from sharks' liver, is rich in vitamin A. But the same vitamin can now be made artificially, without having to kill sharks.

Many COSMETICS, such as anti-wrinkle skin creams and lipsticks, contain oil from sharks.

MORE FASCINATING FACTS

COOKIE CUTTER SHARKS take round bites out of their victims.

During World War II, the Japanese used SHARK LIVER OIL in the engines of their fighter planes.

In the past, sailors attached a SHARK'S TAIL to the bow of their ship to bring them good luck.

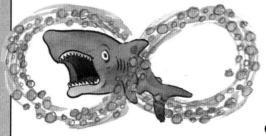

A BASKING SHARK can filter 2,376 gallons (8,994 liters) of water in one hour.

The GREY REEF SHARK swims in a figure eight when it is frightened.

Some AFRICAN TRIBESMEN protected their swords with covers made from shark skin.

On the Pacific island of Samoa, models of sharks were hung from PALM TREES to protect the fruit.

On some Pacific islands, shark teeth are used to make SKIN TATTOOS.

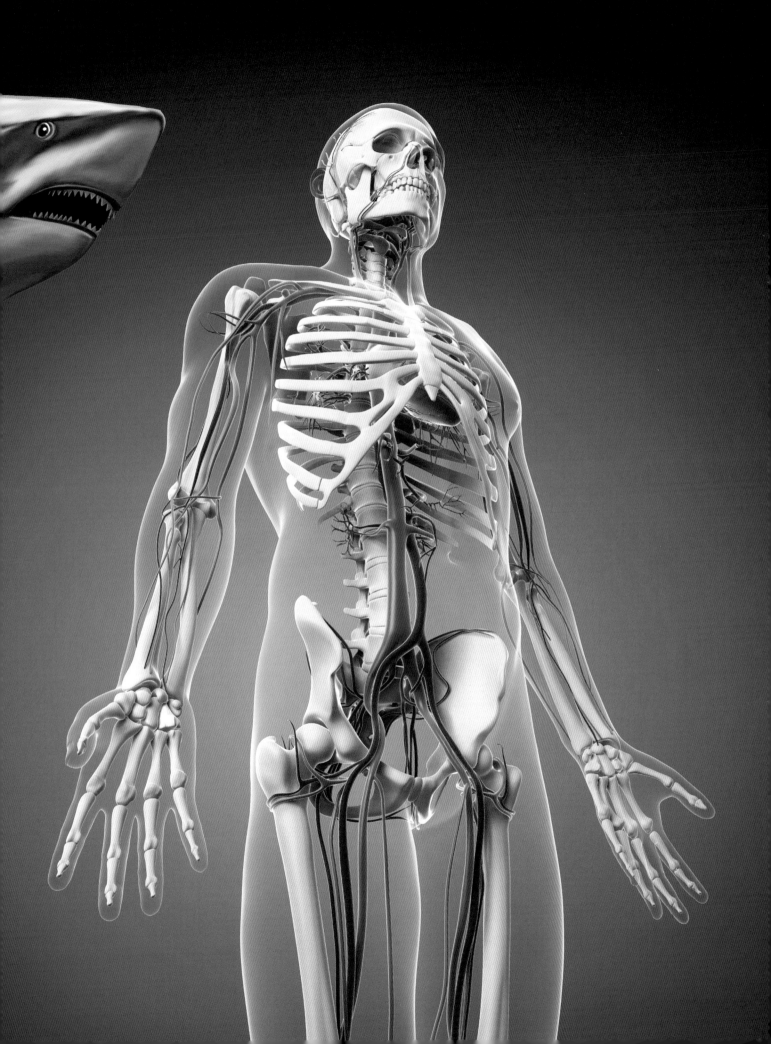

I KNOW ABOUT!

THE AMAZING HUMAN BODY
STRUCTURE AND SYSTEMS 70

DIGESTION 76

HEART AND CIRCULATION 78

SIGHT AND HEARING 84

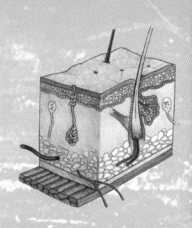

CONTROL CENTER 88

REPRODUCTION 90

MUSCLES AND MOVEMENT 92

STRUCTURE AND SYSTEMS

The human body, and especially the brain that controls it, are extremely complicated and truly amazing. Even so, scientists now understand much of the body's structure, or anatomy, and how the body works. Like all living things, the body is made up of tiny units called "cells." There are many different types of cells in the body, doing different jobs. Cells combine with others of the same type to form tissues. Various tissues combine to form the body's main parts, the organs, and the organs combine to make entire systems.

A HUMAN CELL

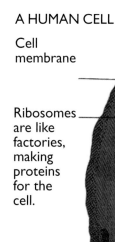

Cell membrane

Ribosomes are like factories, making proteins for the cell.

The nucleus is like a control center.

Golgi bodies are like warehouses, storing proteins.

Mitochondria are like power stations, supplying energy for life processes.

Tissues
There are four basic types of body tissue: skin tissue, muscle tissue, nerve tissue, and connective tissue (above), which includes bone, cartilage, and fat.

Inside a cell
A cell contains even smaller parts known as "organelles," such as ribosomes. Each is like a mini-organ, with a special job to do. The nucleus is the cell's control center and contains the genetic material, DNA (see page 91).

Body organization
The basic building blocks of the human body, and of all other living things, are cells. These microscopic objects are the smallest units capable of the vast range of chemical and physical processes that we call life. There are about 100 trillion cells in the body, of more than 20 main kinds, such as muscle cells and skin cells. Each has a special shape, adapted for the job it does. A group of cells and the substances in which they live make up a tissue.

Ancient beliefs
In ancient times, people had only the vaguest ideas about how the body worked. Traditions and religions often forbade them to cut open, or dissect, the body to study its parts. One of the first great body-scientists was the Roman physician Claudius Galen (AD 129-201). Galen was medical officer for the gladiators who fought bloody battles in the arenas of Ancient Rome. He wrote many books, and his reputation was so great that doctors followed his ideas—even many wrong ones—for 1,500 years afterward.

Organs

An organ is a part of the body with a particular job to do. Some organs are located in body cavities. The heart and lungs are found in the chest cavity, which is separated from the abdominal cavity by the diaphragm. The stomach, intestines, liver, kidneys, and reproductive organs are located in the abdominal cavity, as shown below. Other organs, such as bones and muscles, are found throughout the body.

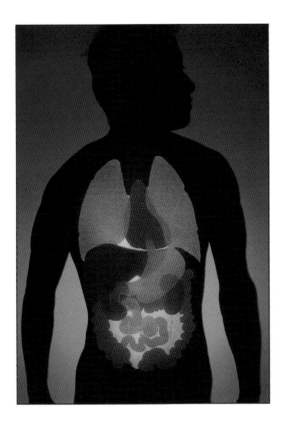

Systems

The organs combine into systems, which are the body's main working units. For example, the lungs and the air passages that lead to them make up the respiratory system (see pages 74-75). The heart, arteries, veins, and other blood vessels, and the blood itself, make up the circulatory system (see pages 78-79). The brain and nerves form the nervous system, which controls body processes. The human body has several organ systems, or main working units, many of which are explored in this book.

Body words

In Ancient Greece and Rome, physicians believed that the parts of the body were made from different mixtures of four fluids, or humors. These were blood (sanguis), phlegm (pituita), black bile (melanchole), and yellow bile (chole). They were connected to the four essences that made up all matter: air, water, fire, and earth. These old beliefs gave rise to many words we still use today. A person in a sad mood is said to be melancholic, because these feelings were thought to be caused by an excess of black bile. Words such as phlegmatic, bilious, and sanguine have similar origins.

Looking into the body

In the 1600s, the microscope was used with great skill by a Dutch scientist, Anton van Leeuwenhoek. By the late 1600s, great anatomists such as Marcello Malpighi were using the microscope to look into the body and discover cells and other tiny parts. Their work laid the foundations of micro-anatomy. Doctors today use scans of various kinds to look inside the body. Unlike ordinary X-rays, which show only bones, these scans show all the soft parts of the body, such as muscles and nerves. A Magnetic Resonance Image (MRI) can reveal the structure of the brain, spine, and facial tissues (below).

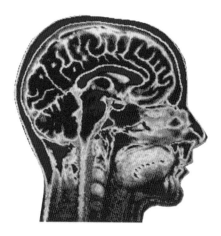

BONES AND JOINTS

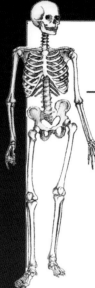

The human body is supported by the hard, strong framework of the skeleton. A single bone is stiff and unbending. But the body can bend and move because many of the bones are linked to each other at flexible, movable joints. Living bones are not dry and brittle, like bones in museums. They are one-third water; the rest is composed of crystals of minerals such as calcium and phosphorus, and a protein called "collagen." Bones have a good supply of blood and nerves and are just as alive and active as the muscles and other soft parts around them.

The skeleton

There are 206 bones in the average human skeleton. A few people have more or less. For example, one person in 20 has an extra pair of ribs, making 13 pairs. Each bone is shaped to protect and support the soft parts around it. The skull, for instance, encloses and protects the brain. There are 30 bones in each arm and hand, and 30 in each leg and foot. The spine, or backbone, is a column of 33 closely-linked bones called "vertebrae." The biggest bones are the femurs, in the thighs. The smallest are the three tiny bones known as the auditory ossicles, in each ear (see page 84). These bones are about the size of rice grains.

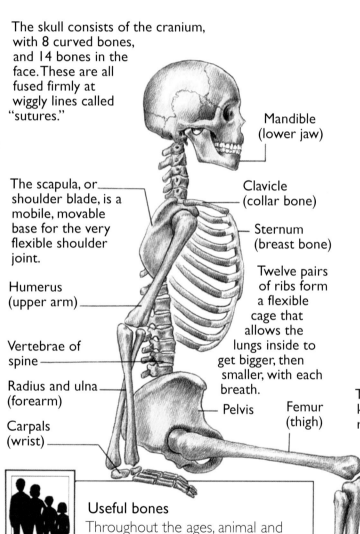

The skull consists of the cranium, with 8 curved bones, and 14 bones in the face. These are all fused firmly at wiggly lines called "sutures."

Mandible (lower jaw)

The scapula, or shoulder blade, is a mobile, movable base for the very flexible shoulder joint.

Clavicle (collar bone)

Sternum (breast bone)

Humerus (upper arm)

Twelve pairs of ribs form a flexible cage that allows the lungs inside to get bigger, then smaller, with each breath.

Vertebrae of spine

Radius and ulna (forearm)

Pelvis

Carpals (wrist)

The patella is the kneecap, a small rounded bone that protects the front of the knee joint.

Femur (thigh)

Tibia and fibula (shin and calf)

Tarsals (ankle)

Phalanges of foot

Useful bones
Throughout the ages, animal and human bones have been used as tools, symbols, ornaments, and decorations in many societies. Animal bones, along with chipped stones, were probably among the first tools used by prehistoric humans.

False color X-ray of hand of a two-year-old boy.

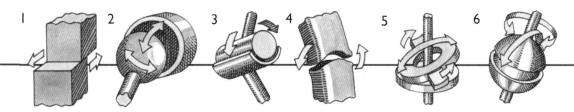

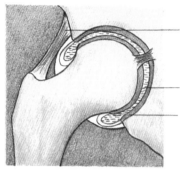

Elastic ligaments hold the bones together.

Joint capsule contains fluid that lubricates the joint.

Smooth cartilage covers the ends of the bones where they touch.

Bones and X-rays

Inside a joint, above, two bones are connected, yet are still able to move. In X-ray photos, bones show up as white shapes. X-rays can detect broken (fractured) bones, or abnormalities as the skeleton grows and matures. Compare the X-ray below, of an adult human hand, with the one on the opposite page. These X-rays illustrate how the bones in the hand and wrist grow and strengthen during childhood.

Types of joints

There are many different kinds of joints in the human body. Among the most important are the plane joint (1) which allows small gliding movements between the vertebrae of the spine, and the ball-and-socket joint (2), found in the hip and shoulder, which allows the limbs to rotate. The hinge joint (3) is found in the knee, ankle, and elbow, and the saddle joint (4) in the base of the thumb. A pivot joint (5) allows the neck to move freely, and ellipsoid joints (6) connect the fingers to the palm of the hand and toes to the sole of the foot.

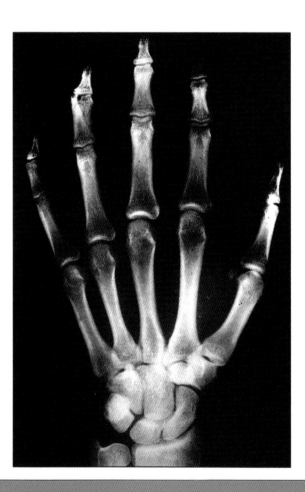

The short and the tall

There are wide variations in the sizes of human bodies, both within one group of people, and between different groups around the world. These variations are based on the sizes of the bones. One of the tallest human groups, the Dinkas, and the smallest, the Pygmies, both come from Africa. Most Pygmies are between 3½ and 4½ feet (1 and 1.4 m) tall. The tendency in developed regions, such as Europe and North America, is for people from all ethnic groups to become taller, due to better health care and diet. A bone expert can tell at a glance whether a skeleton comes from a man or a woman, a large child, or a small adult.

Two adult Pygmies stand beside a Western man of Caucasian origin.

LUNGS AND BREATHING

Nearly all living things, apart from some microbes (tiny organisms), need a supply of oxygen. Oxygen is necessary for the chemical reactions inside a cell known as "cellular respiration." These reactions break up nutrients from food, such as sugars, and set free the energy for driving the cell's life processes. The human body gets its oxygen supplies from the air, which is one-fifth oxygen. We breathe air into our lungs, where oxygen is absorbed into the blood and sent around the body. The nose, throat, trachea, and the two lungs are all involved in breathing; together they are called the "respiratory system."

A new chemical element
Oxygen was discovered to be a chemical element, and vital in breathing, by a series of experiments carried out by scientists such as Joseph Priestley (1733-1804), Carl Wilhelm Scheele (1742-86), and Antoine Lavoisier (1743-94, below). Lavoisier named the substance "oxygen" and discovered it was necessary for burning, and for living things to breathe. This was proved when scientists put animals into glass jars and sucked out the air. This sounds cruel today, since we know that the animals would suffocate and die, but those early experiments helped to explain why life depends on oxygen.

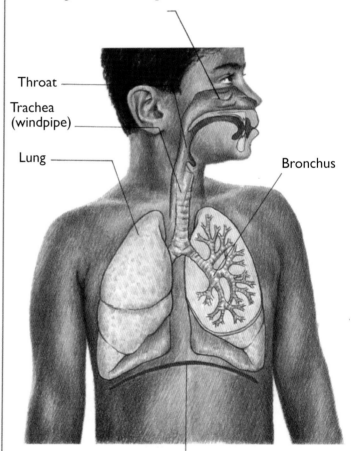

Nasal passage
Breathed-in air flows through the nose, where small hairs filter out bits of dust and floating particles. Sneezing also clears dust particles from the nose.

Throat

Trachea (windpipe)

Lung

Bronchus

Trachea
The trachea divides into two main airways, the bronchi. Their insides are lined with a sticky fluid, mucus, that traps dirt, and tiny hairs that propel the dirty mucus to the top of the trachea where it can be swallowed.

Diaphragm
The chest is filled by the two lungs, the heart, and the main blood vessels. The diaphragm is a dome-shaped muscle sheet that forms the base of the chest and the roof of the abdomen. It is the main breathing muscle.

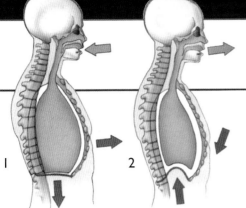

Breathing

Breathing not only obtains vital oxygen, but also gets rid of carbon dioxide. To breathe in, the diaphragm muscle contracts, pulling down the bases of the lungs. The intercostal muscles around the ribs also pull the front of the chest up and out. These movements expand the lungs, sucking in air (1). To breathe out, the diaphragm relaxes. The lungs spring back to their smaller size, puffing out air (2).

Recycled breath

Exhaled or breathed-out air contains less oxygen and more carbon dioxide than normal fresh air. But there is still enough oxygen to keep the body going for a couple more breaths. When first-aiders carry out mouth-to-mouth resuscitation (below), they breathe their own air into the patient's lungs. Great escapologist Harry Houdini (1874-1926) is said to have re-breathed his exhaled air a few times, during the daring escapes he staged underwater.

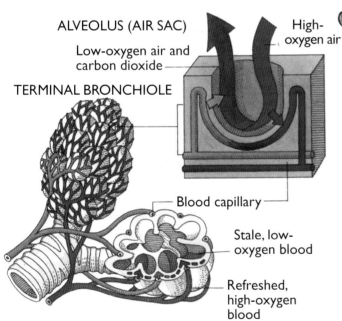

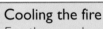

Cooling the fire

For thousands of years, people wondered why we breathe air. The Ancient Greeks, such as Plato and Aristotle, believed that nutrients from food were burned in the heart to make a "vital flame" that brought warmth and life to the body. They thought that the breathed-in air helped to keep the fire controlled.

Inside the lung, the main bronchus (airway) of each lung divides more than 15 times. These airways become smaller, until they form thousands of hair-thin air tubes called "terminal bronchioles." The bronchioles end at microscopic air sacs known as "alveoli." In the alveoli, oxygen seeps from fresh, breathed-in air, through the very thin linings, into the blood flowing through capillaries (tiny blood vessels) on the other side. Blood distributes the oxygen around the body, and carries carbon dioxide back to the lungs, where it is breathed out as stale air.

Exercise is good for you

Aerobic means "with oxygen." Aerobic exercise makes the body's muscles increase their demands for oxygen, because they are using up more energy, thereby making the lungs and heart work harder. In general, this type of exercise improves the health of the lungs, heart, and blood system. In anaerobic (or "without oxygen") exercise, the muscles work for short bursts, without immediately increasing their need for oxygen.

DIGESTION

Energy makes things happen. Every living thing, and every machine, needs energy to power its inner processes. The human body's energy comes from its food. Food also supplies the raw materials and building-blocks for body growth and repair. The set of organs that gets food into the body, breaks it down, digests it, and absorbs it into the blood, is called the "digestive system." It includes the mouth and teeth, esophagus, stomach, and intestines.

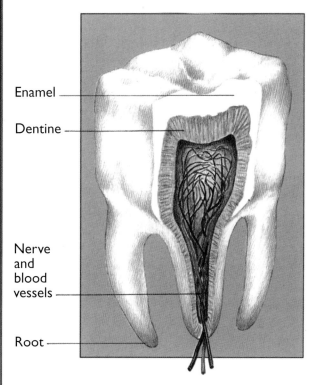

Enamel
Dentine
Nerve and blood vessels
Root

Teeth

Prehistoric people did not have knives and forks. Like wild animals, they used their teeth to cut off and chew mouthfuls of raw food. Today, we cut up and cook our food, but teeth are still important—especially for thorough chewing, which helps the body to digest food more effectively. A tooth is covered with pale enamel, the hardest substance in the body. Under this is dentine, a shock-absorbing barrier. Inside the tooth is the pulp cavity with a good supply of blood vessels and nerves that warn of toothache. Babies grow a first set of milk teeth, 20 in number (in pink above left). From about the age of six years, these fall out naturally, and are gradually replaced by 32 adult, or permanent, teeth. The main types of teeth are shown on the left below.

Grinders
Premolars and molars crush and chew food. An adult has four premolars and six molars in the upper jaw, and the same in the lower jaw, shown here.

Chisels
Incisors at the front of the mouth cut and snip food. An adult has four incisors in the upper jaw and four in the lower jaw.

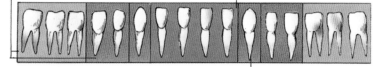

Spears
Canines, next to the incisors, spear and slice food. An adult has two incisors in the upper jaw and two in the lower jaw.

The balanced diet
The science of foods and their effects on the body is called "nutrition." There are many differences in foods around the world. Some people eat mainly potatoes, others rice, and others wheat. Yet all these foods contain lots of carbohydrates for energy. A good, healthy diet contains balanced amounts of the substances shown here, plus vitamins and minerals.

Fiber or roughage for digestive system health
Fruits, leafy vegetables, unrefined breads and cereals, pastas, beans, and lentils.

Moistening food
Three pairs of salivary glands make watery saliva (spit). This moistens the food for chewing. Digestive chemicals known as "enzymes" in saliva start to attack the food and break it down.

Pipe through the chest
The esophagus (gullet) is a muscular tube that pushes swallowed food down through the chest, into the stomach. Normally the gullet is squashed flat by the pressure inside the body.

Acid bath
The stomach contains strong digestive enzymes and also hydrochloric acid, made by its lining. This mixture of powerful chemicals further digests the food for several hours.

Long but narrow
The small intestine contains more enzymes that complete the digestion of food. Nutrients are absorbed through the intestinal lining, into the blood that flows through the lining.

Short but wide
The large intestine forms the undigested and left-over contents into semi-soft masses. It stores these until the body is ready to expel them through the ring of muscle at the end of the tube, the anus.

The digestive system
This system is like a long tube coiling through the body, about 25 feet (7.5 m) long. After the mouth and throat, food goes down the esophagus into the stomach. Hiccupping may occur if we eat or drink too quickly. The stomach is like a storage bag that expands as it fills with a meal, and then squeezes the food into a thick soup. The food passes on to the small intestine, where most nutrients are absorbed. The large intestine absorbs much of the water from the leftovers.

Curious cuisine
Most people in developed regions buy their food from shops. In other regions, people hunt and gather food from the wild. Traditional foods vary around the world. Honeypot ants are a delicacy for Australian Aborigines. Mosquito pie is eaten in parts of Africa. The honeypot ant has sugary fluid in its swollen body. It is a "living pantry" for its nest members.

Fats for growth, repair, healthy nerves, and energy
Dairy, nuts, some plants and their products, and oils, meat, poultry, fish.

Carbohydrates (sugars and starches) for energy
Breads, potatoes, pastas, rice, cereals, fruits.

Proteins for body-building
Meat, fish, eggs, cheese, peas, beans, and other vegetables.

HEART AND CIRCULATION

The body has an efficient transport network to carry supplies and products to where they are needed. It is known as the "circulatory system," because its fluid (blood) circulates, or flows, all around the body, through tubes known as blood vessels. Blood contains three types of cells: red cells carry oxygen, white cells fight disease (see page 83), and platelets cause blood to clot. All these cells float in the liquid part of blood, called "plasma." On its endless journey, blood distributes nutrients, oxygen, and other substances, and collects wastes and by-products.

The blood network

The human body contains between four and six quarts (about 5 liters) of blood, which travels around in an endless one-way circuit. Large, muscular, thick-walled vessels called "arteries" carry blood from the heart. They branch into smaller tubes, which divide further to become capillaries. A capillary's wall is so thin that oxygen and nutrients can easily seep through to the cells beyond, while carbon dioxide and other wastes seep from the cells into the blood. The capillaries join together to make larger tubes, which come together to form large, thin-walled veins. The main veins take blood back to the heart.

The **carotid artery** brings fresh "red" blood from the left side of the heart to the head.

The **jugular vein** returns stale, low-oxygen "blue" blood from the head to the heart.

The **aorta** leads from the left side of the heart, to supply fresh blood to the head, arms, lower body, and legs.

The **vena cavae** bring stale blood from the upper and lower body, to the right side of the heart.

The **pulmonary arteries** take stale blood from the right side of the heart to the lungs, for fresh supplies of oxygen.

The **pulmonary veins** take refreshed blood from the lungs to the left side of the heart.

The **femoral artery and vein** take blood to and from each leg.

Harvey's heart

The English physician William Harvey (1578-1657) realized that the heart was a pump, which forced blood around the body. His discoveries began the modern age of scientific medicine.

Blood groups
In the 1900s, Karl Landsteiner (1868-1943) and other scientists discovered blood groups. They showed that blood from certain groups could not mix safely with blood from other groups. Today, blood transfusions are commonplace and safe, replacing blood lost through injury or an operation. Blood can only be stored for a few months, so it is important for people to keep giving fresh supplies.

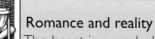

Romance and reality
The heart is a symbol of love and romance. Cards with hearts and romantic messages are exchanged on February 14, the anniversary of St. Valentine, an early Christian martyr (below). Literature is full of references to the heart as the seat of love, courage, and emotions. But the heart is simply a muscular pump, though an amazingly reliable one. True, it does flutter, miss a beat, or speed up during passionate moments. But this is under control from the brain.

The heart has two sides, each with two chambers. The right atrium (upper chamber) receives blood from the main veins.

The left atrium receives blood from the lungs along the pulmonary veins, and pumps it into the left ventricle.

Blood is pumped into the right ventricle. This sends out blood to the lungs to get oxygen.

Aorta

The muscular heart wall is thickest in the left ventricle. This chamber must pump out blood through the aorta to the whole body.

The pulse
When the body is active, the muscles need more oxygen and nutrients. So the heart pumps faster and harder. With each beat, a bulging pressure wave of blood passes out along the arteries. This is known as the "pulse." When you are resting, your pulse rate is about 70-75 beats per minute. After strenuous exercise, your pulse rate may more than double.

The heartbeat
When you are at rest, the heart squeezes once a second. Blood oozes into the atria (1), and is pumped through the heart valves into the ventricles (2). The ventricles' muscular walls contract to pump out the blood (3). The cycle starts again (4).

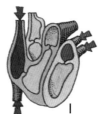 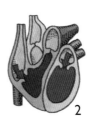 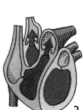 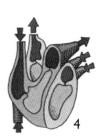

1　2　3　4

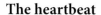

Fingers on the pulse
You can hear a person's heartbeat by placing your ear on his or her chest, slightly to the left. Find the pulse of the radial artery in the inside of the wrist, or of the carotid artery in the neck under the side of the jaw. Feel gently with your fingers, not your thumb, since it has its own small pulse that may cause confusion.

WASTES AND HORMONES

The body's life processes produce dozens of wastes and by-products. These are collected by the endlessly circulating blood, and dealt with mainly by the liver and kidneys, the chief organs of the excretory (waste-disposal) system. In addition, blood circulates special body chemicals called "hormones." These control various internal processes such as digestion, fluid balance, growth, and maturing from a child into an adult woman or man.

Processing foods
The liver is the largest organ in the body. It has a rich blood supply from the stomach and intestines, and receives digested nutrients. It makes toxic (poisonous) substances safe, stores some vitamins and carbohydrates, and converts nutrients into useful forms.

Filtering blood
The kidneys are blood filters. Inside each kidney, one million microscopic filter units called "nephrons" take waste products, such as urea, from the blood, along with excess water. These wastes-and-water form the liquid *urine*, which trickles down the ureters to the bladder.

Ureter

Doing two jobs
The pancreas has two distinct roles. One is to make strong digestive enzymes. These pour along the pancreatic duct into the small intestine to help digest food. The other role is to produce hormones such as insulin.

Urethra

Testing urine
During the Middle Ages, many physicians believed that a patient's urine could show exactly what was wrong with him or her. The urine in a glass tube was held up to the light and examined carefully. This process, called urinoscopy, continued for several centuries until physicians realized it was not very accurate. Today, urine is still examined and used to detect diabetes and other diseases, but this is usually done by chemical tests in the laboratory.

Dealing with wastes
The body makes several types of wastes. Digestive left-over wastes are expelled through the anus. Some blood by-products are processed and altered in the liver. The unwanted substance carbon dioxide is breathed out through the lungs. The main wastes in the blood are removed as a fluid, urine, by the urinary system. This system is made up of the two kidneys, the bladder, the ureter tubes, which link kidneys and bladder, and the urethra, which empties the bladder to the outside.

Storing urine
The bladder is a balloon-like storage bag for urine. When it holds about one cup, stretch sensors in its wall send messages along nerves to the brain, signaling that it needs emptying.

Master gland
The pituitary, just under the brain, makes several hormones and similar chemicals that control other hormonal glands in the body.

Energy regulator
The thyroid gland is shaped like a bow tie in the lower neck. It makes "thyroxine," which affects processes such as the way the body uses energy.

The long and the short of it
Stories about very tall or short people exist in many cultures. Such stories are often exaggerated into legends about giants or tiny humans. In real life, adult humans have grown to more than 8 feet (2.4 m) and less than 2 feet (0.6 m) in height. Humans may have been even bigger or smaller in the distant past. Such extremes are often due to too much or too little growth hormone, made by the pituitary gland.

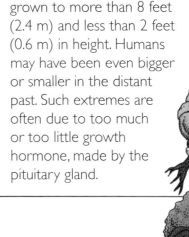

Insulin factory
The pancreas produces the hormone insulin, which controls the way cells take in and use energy-containing sugars.

Chemical control
Hormones are the body's chemical messengers. Each is made in a certain gland. Hormones circulate in the blood and affect specific cells and tissues, making them work faster or slower, or releasing their products. Along with the brain and nerves, hormones help the body to function as a coordinated whole.

Female sex glands
In a woman, the ovaries produce the eggs that can grow into babies if fertilized. They also produce hormones that help to control the process of maturing from girl to woman, and the monthly menstrual cycle.

Male sex glands
In a man, the testes are small, egg-shaped glands hanging below the abdomen in a skin bag, the scrotum. They produce the sperm for reproduction. They also produce hormones that help to control sexual characteristics and the maturing from boy to man.

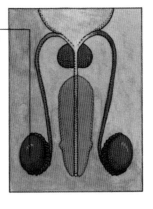

Adrenal glands and kidneys
The adrenal gland on top of each kidney makes a hormone that affects the balance of body water and salts, and adrenalin, that prepares the body for emergency action. The kidneys produce renin, which regulates fluid content and blood pressure.

BODY DEFENSES

The world is full of germs. They are in the air we breathe, in the soil, on plants and animals, even on our skin, and in some of the food we eat. Some are harmful microbes, such as the bacteria and viruses that cause diseases. The body has several lines of defense against them, including the outer covering of skin, and two systems with scattered parts: the lymphatic and immune systems. In some cases, once the body has defeated a type of germ, it is always protected, or immune, against it.

Lymphatic and immune systems

Lymph is the body's "other blood." Lymph fluid collects in the spaces between cells, is channeled into tubes called "lymph ducts," and flows through lymph nodes, where it is filtered and cleaned. Traveling around the body, lymph distributes fats and digested nutrients, and collects wastes, which are returned to the blood and filtered out by the kidneys. Lymph, and the blood itself, carry the germ-fighting white blood cells that form the basis of the immune system. These cells recognize germs and make chemicals called "antibodies" that kill them.

The lymph network

The lymphatic system is body-wide, like the circulatory system. There are collections of bean-sized lymph nodes in the armpits, groin, and other places. The lymph ducts finally come together into large tubes that empty into main veins, near the heart.

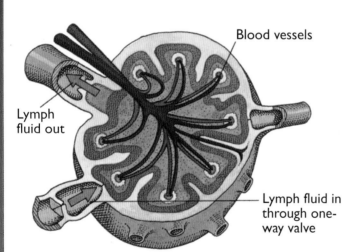

Lymph node (left)

Lymph fluid flows slowly through the node, where cells are added, broken down, and recycled according to the body's needs. In particular, white cells from the blood are stored and multiply, ready to fight germs.

Vaccination

Blood and lymph contain billions of germ-killing white blood cells. The body can be tricked into defending itself by injecting it with disabled versions of real germs. These prepare the body so that if real germs invade, they are killed at once. This process, vaccination, was introduced by a British doctor, Edward Jenner (1749-1823, right). In 1796, Jenner successfully vaccinated a patient against the disease smallpox.

Deadly rhyme

"Ring-around-a-rosy, A pocket full of posies, Ashes, to Ashes, We all fall down." This nursery rhyme originated at the time of the Great Plague, which swept Britain in 1665. As the plague spread, people carried posies, hoping the scent would prevent the disease. If they fell ill, one of the first signs was sneezing, and death would follow. Plague pits were dug (right) to dispose of the bodies and to try to prevent further spread of the disease.

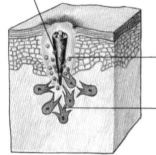
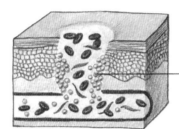

Sealing wounds

Small cuts and wounds happen all the time, inside the body and outside on the skin. At a wound, no matter how small, blood undergoes a chain of chemical changes. If the skin is punctured, for example, by a splinter (above), the blood becomes thicker and sticky, and white cells gather to attack any germs that may try to get in (1).

As the splinter is removed, tiny cell fragments that normally float in the blood, known as platelets, become stuck in the stickiness and make a gooey barrier. This process is known as "coagulating" or "clotting" (2). The clot hardens and seals the wound to stop blood and body fluids from leaking out, and to prevent dirt and germs from getting in (3). Gradually the skin or other tissue grows back and heals itself.

Healers through the ages

Different societies have many types of healers or shamans. In prehistory, shamans may have worn the skins of animals during ceremonies (left). Modern medicine relies on factory-produced drugs, chemical tests, advanced surgery, and complicated equipment. There is no doubt that it has been amazingly successful. But there has been renewed interest in traditional medicines that use natural products and place more importance on the mind and soul.

AIDS

One of today's greatest health challenges is AIDS, Acquired Immune Deficiency Syndrome. A microbe called HIV (Human Immunodeficiency Virus) infects the body and attacks the immune system, so the body cannot protect itself against other microbes. Millions of people around the world have HIV. Drugs slow the course of this fatal disease, but no cure has been discovered yet.

A human cell infected with HIV

SIGHT AND HEARING

An animal needs to sense the world around it in order to find food and drink, locate shelter and mates, and detect and move away from danger. The organs specialized to do this form the sensory system. They are closely connected to the nervous system. The human body's five main senses are sight, hearing, smell, taste, and touch. Our sense of balance is also vital. This involves the semicircular canals in the ears, and information from the skin about touch, from stretch detectors in the muscles in joints about body position, and from the eyes about head position.

Inside the ear
The ear detects patterns of sound waves traveling through the air. The outer ear flap funnels sound waves into a one-inch-long tube, the outer ear canal. At the end of this tube is the eardrum, which vibrates as sound waves bounce off it. The vibrations pass along a chain of three tiny bones, the ear ossicles, and into the fluid inside the cochlea. There, the vibrations are transformed into electrical signals known as "nerve impulses," which flash along nerves to the brain.

Eustachian tube
The Eustachian tube runs between your ear and throat. It opens during swallowing and yawning, to equalize air pressure within the ear.

Sense of balance
The semicircular canals are involved in balance. They are filled with fluid. As the head tilts, tiny hair cells in the fluid move, sending signals to the brain.

Outer ear flap

Ear ossicles
Hammer (malleus)
Anvil (incus)
Stirrup (stapes)

Outer ear canal

The cochlea
A delicate membrane inside the cochlea is covered with microscopic cells that have hairs sticking from them. The vibrating fluid in the cochlea shakes the hairs and generates nerve impulses from the cells.

Eustachian tube to throat

The eardrum
This disk of tightly-stretched, skin-type tissue is about as big as the nail on the little finger. Like skin, it can heal itself if slightly cut or torn.

Hard of hearing
People who are deaf (unable to hear) have been at a great disadvantage in the past. Ear trumpets that collect sound waves and concentrate them into the ear have been used for centuries. A Victorian version is shown here. Today, deaf people can learn lip-reading, and use sign language with their hands. Modern electronic hearing aids can also help some deaf people.

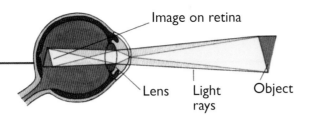

Inside the eye

The eye is a sphere about one inch across, filled with clear jelly. It is specialized to detect patterns of light rays and transform them into tiny electrical signals, nerve impulses, that flash along nerves to the brain. At the front of the eye is a clear domed part, the cornea. Behind this is a colored ring, the iris, with a hole in the center, the pupil. Behind the pupil is a clear bulging disk, the lens, which bends the light rays. Light rays pass through the cornea, pupil, and lens, into the interior of the eye. They shine on a very thin layer lining the inside, the retina. This contains more than 120 million cells, known as rods and cones, because of their shape. The rods and cones turn the colors, patterns, and brightnesses of light-ray energy into nerve-signal energy, and send the nerve impulses along the optic nerve to the brain for analysis.

The world upside down?

Because of the way a lens works, the image it shines onto the retina is upside down. But a baby's brain has no concept of up or down. It learns to turn the images "right side up" almost from the very beginning, as a natural part of sight.

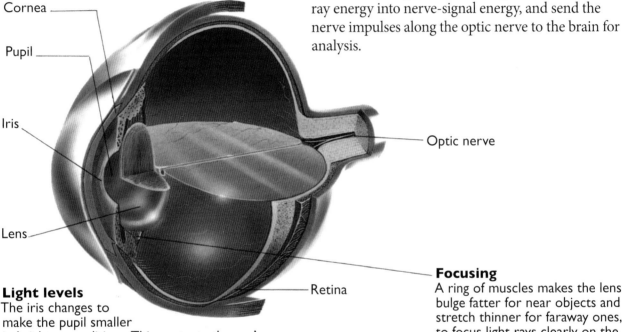

Light levels

The iris changes to make the pupil smaller in brighter conditions. This protects the eye's delicate interior from being damaged by too much light.

Focusing

A ring of muscles makes the lens bulge fatter for near objects and stretch thinner for faraway ones, to focus light rays clearly on the retina.

Better sight

Eye glasses and contact lenses change the path of light rays, so that people whose eyes cannot focus properly are able to see clearly and sharply. The first eye glasses were in use by the 13th century. The system of Braille allows blind people to read, by running their fingers over tiny patterns of raised dots on the page. It was invented in 1824 by Louis Braille (1809-1852), a blind French student, when he was 15.

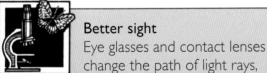

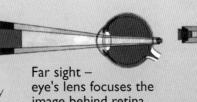

Far sight – eye's lens focuses the image behind retina.

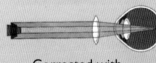

Corrected with convex lens.

Near sight – lens focuses the image in ront of retina.

Corrected with concave lens.

SMELL, TASTE, TOUCH

The human body's main senses are sight and hearing. However, our sense of smell is also important, helping us to identify things at a distance, unlike taste and touch. Taste is similar to smell, but detects flavor substances in foods and drinks, rather than odor substances floating through the air. Both these senses check that foods and drinks are not bad or rotten, as revealed by unpleasant odors and flavors. The skin also has a warning function, signaling if the body is in danger from extreme heat, cold, or pressure.

Inside the nose and mouth

Air, carrying odor substances, passes through the nasal cavity as you breathe in and out. The odors land on a batch of toothbrush-like olfactory hairs in the roof of the cavity. Microscopic cells attached to the hairs send nerve impulses to the brain when they detect certain odors. Similarly, flavor substances from foods are detected by tiny, onion-shaped groups of cells known as "taste buds." These are scattered over the tongue, among the larger lumps and bumps known as "papillae." The taste bud cells send nerve impulses to the brain when they recognize four main flavors: sweet, salty, sour, and bitter.

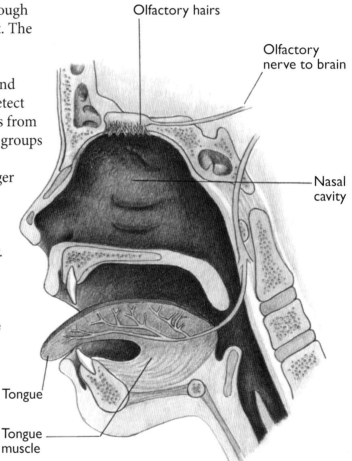

Testing smell and taste

Test your senses of smell and taste by trying this experiment with a friend. Prepare small samples of sweet, sour, salty, and bitter foods. See if you can tell which sample is which when you are blindfolded. Then, try the experiment again, holding your nose. How successful are you this time at detecting different tastes?

Taste buds

An average person has about 10,000 taste buds. They are located mainly on the tongue, among the tiny projections known as "fungi-form papillae" that appear in pink in the magnified false-color image on the left, above. Conical papillae (blue) form a rough surface that helps in the chewing of food. There are also taste buds on the insides of the cheeks, on the roof of the mouth, and on the upper throat.

Sweaty lies
When people are nervous, they tend to sweat. This reduces the skin's resistance to small, harmless amounts of electricity. The polygraph or "lie detector" measures the electrical resistance of the skin. It has been used to test whether people are calm, and probably innocent, when questioned, or nervous, and perhaps guilty.

Skin color
Groups of people who have traditionally lived in certain parts of the world have different skin colors. Where the sun's rays are strong, skin makes more of its natural dark coloring substance, melanin. This protects the tissues beneath from the sun's potentially harmful UV (ultra-violet) rays. Over thousands of years, evolution has produced dark skin, compared to the light-colored skin of people who live in temperate regions.

American Indian   European   Asian   African

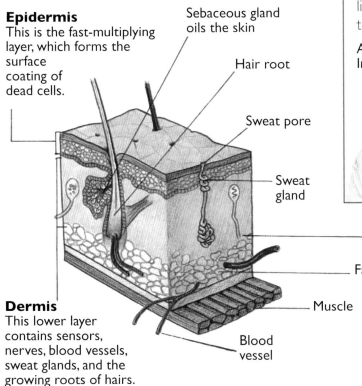

Epidermis
This is the fast-multiplying layer, which forms the surface coating of dead cells.

Sebaceous gland oils the skin

Hair root

Sweat pore

Sweat gland

Light touch receptor (Meissner's corpuscle)

Fat layer

Muscle

Blood vessel

Dermis
This lower layer contains sensors, nerves, blood vessels, sweat glands, and the growing roots of hairs.

Sensors
In addition to the cold and light touch receptors shown, the skin contains sensors for heat, pain, and pressure.

Washing is fashionable
Thorough personal hygiene, with daily washes, is relatively recent. In the past, the nearest running water was the local river. Rich people used perfumes to mask their body odors. By the last century, people knew that some germs cause disease. They began to wash more often, sometimes helped by a maid (right).

The skin
The skin is the body's supple, protective coat. It helps to keep out too much heat, cold, water, germs, and harmful rays, and keep in body warmth and fluids. Its surface consists of microscopic skin cells, dead and hardened with the substance *keratin*. Every day, millions of these cells are worn away as the body moves, wears clothes, and washes. They are replaced by cells below multiplying and moving upward. Under this layer are microscopic sensors that detect touch, pressure, pain, heat, and cold. Under the dermis is a layer of fat, which absorbs shocks and helps insulate the body from heat and cold.

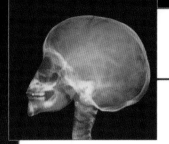

CONTROL CENTER

The brain is the control center of the body, and the place where consciousness and the mind are based. Some 100 billion nerve cells, each one connected to up to half a million others, make up an unimaginably complicated network for nerve impulses. Much of the brain's activity does not take place consciously. The "automatic brain" controls heartbeat, breathing, digestion, temperature, and many other body processes, without your conscious involvement. Certain areas of the brain are concerned with specific functions.

Thalamus
This egg-shaped part is a relay station for sensory input. Most sensory nerves pass into the thalamus, which redirects impulses, so that the "thinking" cerebral cortex can concentrate on what is important.

Hypothalamus
Small as a fingertip, the hypothalamus regulates basic urges and desires such as hunger, thirst, body temperature, and sexual activity.

Cerebral Cortex
This wrinkled sheet of "gray matter," about 1/8 inch thick, is the outer wrapping for the two cerebral hemispheres that make up 90 percent of the brain. The cortex deals with thoughts, memories, learning, sensations, and body control (see opposite).

Cerebellum
The wrinkled lobes of the cerebellum look like a mini version of the cerebral hemispheres. They coordinate and organize muscle actions.

Medulla
The lowest part of the brain narrows and merges with the spinal cord. It is involved in automatic processes such as pulse rate, digestion, breathing, and chewing.

Reading head bumps
In the 19th century, phrenology was a popular and important "science." A phrenologist was an expert at reading the lumps and bumps on the human skull, which were supposed to reveal the person's intelligence, friendliness, reliability, truthfulness, emotional behavior, and other aspects of the personality. Phrenology is now recognized as fake by modern scientists.

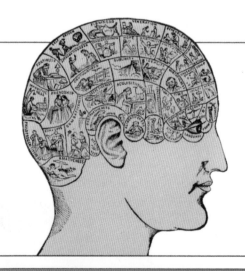

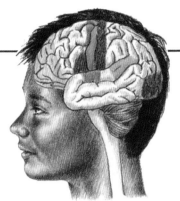

Brain areas
Different areas of the brain's cerebral cortex are the centers of various kinds of mental activity.

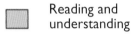 Reading and understanding

Hearing

Speech

Body motor center

Awareness

Visual

Body sensory center

"You are getting sleepy..."
One mysterious but scientifically recognized feature of the mind-and-brain is hypnosis. This is also called "mesmerism" after one of its pioneers, Austrian physician Franz Mesmer (1734-1815). The hypnotized person will do many things asked of him or her and can be "put to sleep." Hypnotism has become a useful technique in treating various mental disorders, easing pain, and controlling obsessions and habits, such as gambling and smoking.

Animal brains
Humans are the most intelligent animals on Earth. This is due partly to our large brains, and especially the area of the cerebral cortex. More accurately, it is linked to the size of the brain and its cortex in proportion to the whole body. In general, animals that we consider to be "intelligent" have large brains in relation to their body sizes. The average human brain weighs 2½-3 pounds (1.13-1.36 kg). In various mammals, different proportions of the cerebrum are centers of hearing, seeing, and so on, depending on how important those functions are to the animal. Smell is a very important sense for rats and ground shrews, but it is less important to chimpanzees. Chimpanzees and other intelligent primates have a much larger proportion of non-specific brain.

Rat 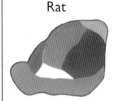   Chimpanzee    Ground shrew

Motor   Hearing   Seeing
Smell   Sensory   Non-specific

Head-hunters
Until recent times, some groups of people used to sever and preserve the heads of their enemies to commemorate victory in battle. Some have even eaten their enemies' brains in an attempt to gain extra courage and wisdom. These traditions were carried out in parts of South America, Papua New Guinea, and Indonesia, shown in the illustration on the right. They are now exceptionally rare.

REPRODUCTION

The reproductive systems of women and men are the parts specialized for producing babies. Mating (sexual intercourse) and pregnancy are a natural part of the life cycle of all mammals. Parents have young, who grow up and have young of their own, and so on. The biological basics of human reproduction are much the same as in our mammal relatives, such as monkeys.

MALE REPRODUCTIVE ORGANS

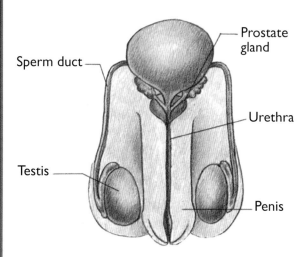

FEMALE REPRODUCTIVE ORGANS

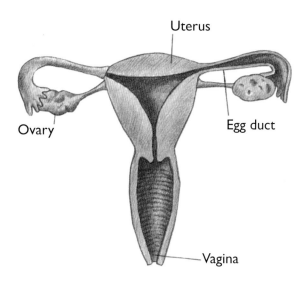

In the male, the testes are glands that make millions of sperm cells. During sexual intercourse, the man's penis becomes longer and stiffer, and is inserted into the woman's vagina. Sperm come along the urethra and enter the vagina. They swim through the uterus and into the egg ducts, toward a ripe egg.

About every four weeks, one of the woman's ovaries releases a tiny, ripe egg cell that travels down the egg duct towards the uterus. The uterus is ready to nourish the fertilized egg. If the egg is not fertilized by sperm, the uterus lining breaks down and flows out of the vagina in menstruation, the monthly blood loss, or period.

Families big and small

In times gone by, and in some countries today, parents have lots of children. Often, as soon as they are able, the children must work to bring back money for the family. In other parts of the world, parents have only one or two children, and may practice birth control (contraception) to prevent the possibility of having any more. This may be due to personal preference, or lack of money and jobs, or because of their countries' official encouragement of small families.

Only one sperm cell fuses with the egg cell to fertilize it.

Fertilization
An egg cell cannot develop into a baby unless it is fertilized by a sperm cell from the male. The head of one sperm merges into the egg cell (above), and the genetic material (DNA) of each cell come together. The DNAs contain all the instructions, in the form of a chemical code, to make a new living, breathing human body.

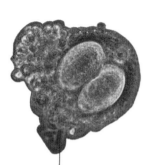

The egg cell begins to divide 24 hours after fertilization.

The secret of life
In 1953, American biologist James Watson (1928-present) and English biochemist Francis Crick (1916-2004) worked out the structure of DNA. Its shape is a double helix, like two intertwined spiral staircases. The strands can separate and copy themselves, as cells divide. The sets of DNA are in the nucleus of the cell, tightly coiled into X-shaped structures called "chromosomes." Watson and Crick's discovery explained how general features are passed on from parents to children (right), yet each child (apart from identical twins) is always slightly different.

Embryo at six weeks

Fetus at four months, with the umbilical cord that connects it to the mother's placenta

Pregnancy
After fertilization, the egg cell divides, forming a pinhead-sized ball of a hundred or more cells. This settles into the nourishing, blood-rich lining of the uterus, and its cells continue to multiply. Gradually, the cells become specialized into nerve cells, muscle cells, blood cells, and so on, and a tiny baby takes shape. In this stage the baby is called the "embryo." Two months after fertilization, the baby is about one inch long. From then until birth, at nine months after fertilization, it is known as a "fetus." At birth, the baby passes through the neck, or cervix, of the uterus, through the vagina, to the outside world.

The placenta
The baby is nourished by a special organ called the "placenta." This passes nutrients and oxygen from the mother's blood across a thin membrane to the baby's blood, through a lifeline called the "umbilical cord."

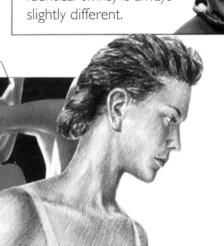

- Fetus
- Uterus
- Closed cervix of uterus

MUSCLES AND MOVEMENT

All the body's movements are powered by muscles. Muscle tissue is specialized to contract, or get shorter. The body has three main kinds of muscles. One is skeletal muscles, attached to the bones of the skeleton, which you use to move about. There are more than 600 skeletal muscles, from the huge gluteus in the buttock to tiny finger and toe muscles. The other kinds of muscles are cardiac muscle in the heart (top left above) and smooth muscles in the stomach, intestines, and other internal organs (left above).

Inside a muscle

A skeletal muscle has a bulging central part known as the "body." This tapers at each end into a rope-like tendon, which anchors the muscle to a bone. As the muscle contracts, the tendons pull on their bones and move the body. The muscle body is divided into bundles of hair-fine fibers called "myofibers." These long cells contain proteins that slide past one another to make the cell shorter in length.

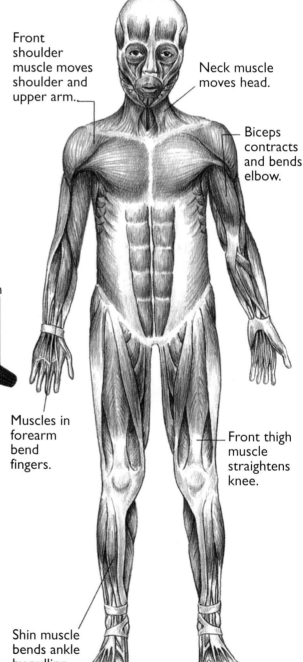

Front shoulder muscle moves shoulder and upper arm.

Neck muscle moves head.

Biceps contracts and bends elbow.

Muscles in forearm bend fingers.

Front thigh muscle straightens knee.

Shin muscle bends ankle by pulling up foot.

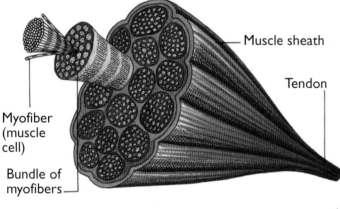

Muscle sheath

Tendon

Myofiber (muscle cell)

Bundle of myofibers

Changing fashions

Bulging muscles have been in and out of fashion through the centuries. A few hundred years ago, plump bodies were seen as desirable. Today, some men and women like to look slim. Other people work hard at body-building, training and lifting weights in the gym. They strive to increase the thickness of their muscle fibers through special exercises and diet.

Stories of the strong

Legends from many different cultures tell of well-muscled, strong men and women. Some are heroes, others are villains. Hercules of Ancient Greece had to undertake 12 "herculean" (very difficult) tasks or labors. In the Bible, the boy David fought and killed the giant Goliath with his slingshot. Samson was a hero who fought the Philistines, but he lost his strength when Delilah tricked him into having his hair shorn. Blinded and chained, he pushed the columns of the Gaza Temple and brought it crashing down on himself and his captors, as pictured right.

Triceps contracts, straightening elbow. Biceps relaxes and stretches.

Master of art and science

During the Renaissance period, from about the 14th century, there was a rebirth of fascination in the beauty of the human form, and a scientific interest in the structure and workings of the body. Foremost in this field was the genius of art and science, Leonardo da Vinci (1452-1519). He performed amazing dissections of the body, especially the muscle system, and drew them with unparalled skill and mastery, as shown here.

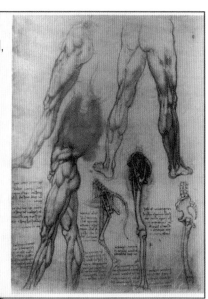

Biceps contracts, bending elbow. Triceps relaxes.

Quadriceps contracts, straightening knee.

Hamstring muscles relax and stretch.

Quadriceps and hamstrings tensed to maintain crouched position.

Muscle pairs

A muscle contracts to pull on its bone. But it cannot do the reverse–actively get longer and push the bone the other way. So many of the body's muscles are arranged in opposing pairs, attached across the same joint. One partner of the pair pulls the bones one way, bending the joint. The other pulls the other way and straightens the joint, while its partner relaxes. Even a simple movement also involves many other muscles that keep the body balanced.

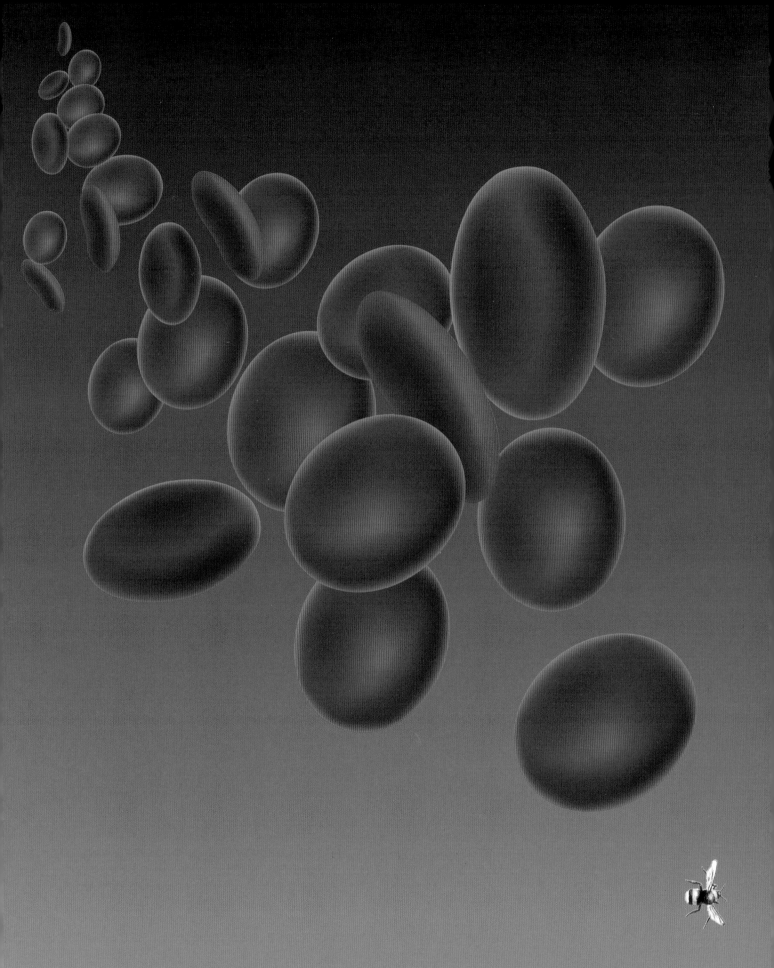

Photo credits:
Robert Harding Picture Library; Roger Vlitos; Frank Spooner Pictures; Mary
Evans Picture Library; Hulton Deutsch; Science
Photo Library; Spectrum Colour Library

I KNOW ABOUT! INSECTS

WORLD OF WONDER

LIFE CYCLES 100

BEETLES 110

106 POISONS AND STINGS

BUTTERFLIES AND MOTHS 114

BEES AND WASPS 116

ANTS AND TERMITES 118

CLASSIFICATION OF INSECTS 121

WHAT ARE INSECTS?

Insects are the most successful of all animal groups, making up 85 percent of the whole animal kingdom. There are as many as 10,000 insects living on every square yard of the Earth's surface. There are many different kinds of insects, but all share a common body design, adapted to cope with every possible environment, and to eat every possible kind of food. All adult insects have a segmented body, which is divided into three parts: head, thorax, and abdomen.

An insect's skin is made of a tough substance called "chitin." This forms a hard shell, or exoskeleton, which protects the insect's organs. The leg and wing muscles are securely anchored to the exoskeleton. It is waterproof, and prevents the insect from drying out. But it does not allow air through. Holes in the skin, called "spiracles," lead to breathing tubes. The exoskeleton does not grow. As an insect gets bigger, it must shed its old skin, and grow a new one. The outer skin, or cuticle, is patterned and colored for camouflage or warning (see pages 106-107).

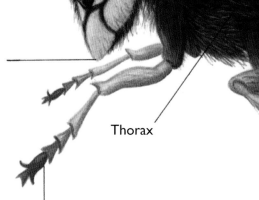

Antennae (see pages 104-105)

Compound eye (see pages 104-105)

Mouthparts (see pages 104-105)

Thorax

Six jointed legs (see pages 102-103)

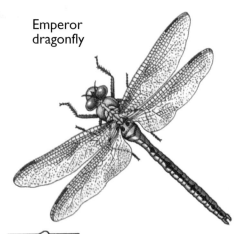

Emperor dragonfly

Common cockroach

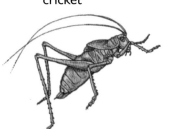

Bush cricket

Firebug

Preserved in stone

Insects first appeared on Earth about 370 million years ago. Early species had no wings; they fed on the sap and spores of the newly-evolved land plants. Insects were the first creatures to conquer the air, 150 million years before birds first flew. This is a fossil of an early dragonfly that lived 300 million years ago, in the steamy Carboniferous forests with the ancestors of the dinosaurs.

All insects have three pairs of jointed legs, and most have four wings. Insects from some easily recognizable insect groups are shown above. The classification of insects is explained on page 121.

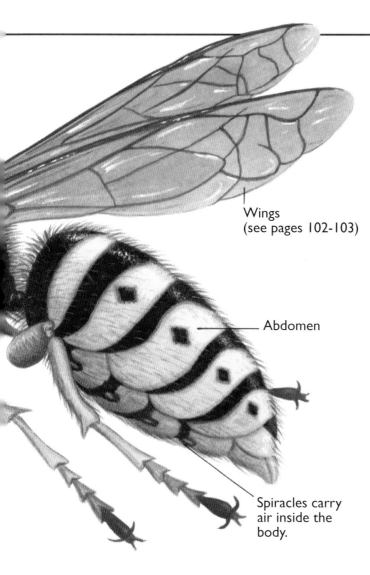

Wings (see pages 102-103)

Abdomen

Spiracles carry air inside the body.

Biblical insects

The Bible contains many stories about Samson, a hero who possessed great strength. One story tells how he killed a young lion that threatened him. Later, he noticed bees flying from the lion's body and found honey inside it. He made his discovery into a riddle: "Out of the eater came forth meat, and out of the strong came forth sweetness." No one could guess the answer. In fact, the insects were probably not bees at all, but carrion flies that live on rotting flesh. The explanation of the honey is still a mystery!

Samson discovers the bees.

The head contains a simple brain that receives messages from the sense organs and controls the muscles. The thorax is made of three segments fused together. It carries the legs and wings. The abdomen contains the organs for digestion and reproduction.

Seven-spotted ladybird (beetle)

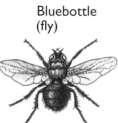

Bluebottle (fly)

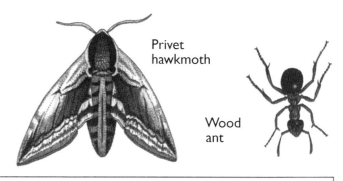

Privet hawkmoth

Wood ant

Insect relatives

Insects belong to a group of animals called "arthropods." They all have segmented bodies with hard exoskeletons. But the other arthropods pictured here are not insects. Spiders have eight legs. Their body segments are fused in two parts—a head-thorax and an abdomen. Millipedes and centipedes have many body segments, with legs on each.

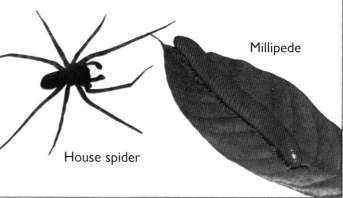

Millipede

House spider

LIFE CYCLES

Insects hatch from eggs. Young insects are eating machines. They consume as much food as they can in order to grow to adulthood as quickly as possible, and shed or molt their skins as they expand. Their goal as adult insects is to mate and lay eggs, and so the life cycle begins again. The change from young to adult form is called "metamorphosis." In many species, the young insects turn into adults gradually. In some species, however, the change is quite dramatic. These young insects, or larvae, often have a different diet than their parents. They live their early life in a form very different than their adult shape.

Egg

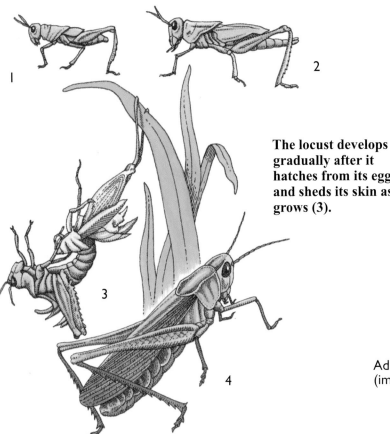

The locust develops gradually after it hatches from its egg, and sheds its skin as it grows (3).

Adult (imago)

Incomplete metamorphosis

In many species, the transformation from young to adult form is gradual. Young locusts are called "nymphs." Newly hatched nymphs look a little like small, wingless adults. As they grow, they shed their skin whenever it gets too small. Each time they molt, they become more like adults. They emerge from their final molt complete with wings and reproductive organs, ready to fly off, find a mate, and lay eggs. This gradual change is *incomplete metamorphosis*. Cockroaches, bugs, and dragonflies also have incomplete metamorphosis.

Caring parents

Many insect parents go to a lot of trouble to protect their newly-hatched young and provide a ready supply of food. Butterflies lay their eggs on a suitable food plant. Gall wasps choose plants that will grow a protective gall around the larvae. Some wasps paralyze an insect victim and drag it into their burrow for the larvae to feed on. Female cockroaches carry their eggs around with them until they are almost ready to hatch. Earwig mothers carefully clean their eggs and young.

Caterpillar watch (3 months)
Summer is the time to observe the life cycle of butterflies, such as the large white butterfly. You may be lucky enough to find some eggs or caterpillars, and wish to remove them for study and identification. Place them, with the leaves on which you found them, in a box covered with muslin. Keep the box in a cool, moist place and replace the leaves with the same kind every day. Watch the caterpillars grow, and count how many times they molt before they pupate. When the butterflies emerge, watch them unfurl their wings by pumping blood into the veins. Let the butterflies go.

The red admiral, like all butterflies, has four very distinct stages in its life cycle, as egg, caterpillar, chrysalis, and finally adult.

Caterpillar (larva)

Chrysalis (pupa)

Complete metamorphosis
Some kinds of insects undergo a much more dramatic change as they grow from larvae to adults. As grubs or caterpillars, the young of these species look nothing like adult insects. They eat constantly and molt whenever their skin gets too small. Eventually, the larvae anchor themselves to a safe spot by a silken thread, and molt to reveal a chrysalis. The chrysalis looks motionless, but inside, a great transformation is going on. Long, jointed legs and antennae form, and wings develop. At last the chrysalis bursts open and a completely different animal emerges. This dramatic change is called "complete metamorphosis."

Life charts
The red admiral spends one week as an egg, five weeks as a caterpillar, two weeks as a pupa, and nine months (39 weeks) as an adult. The pie chart below represents this life cycle. Make a similar chart for the stag beetle, which spends two weeks as an egg, three years (156 weeks) as a caterpillar, eight months (35 weeks) as a pupa, and four weeks as an adult. Add the total number of weeks. Figure out what percentage each stage is by multiplying each by 100 and dividing by the total of weeks. Then, multiply each figure by 3.6 to find out how many degrees it represents of a circle. Mark out the degrees with a protractor.

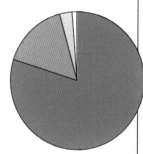

Some species of ichneumon wasps lay their eggs in a living caterpillar. Once hatched, the grubs feed on the caterpillar's insides. The grown larvae burst through the caterpillar's skin and turn into pupae.

GETTING ABOUT

Insects are masters of land, air, and fresh water. They use their legs and wings to move about efficiently in each environment. Most kinds of insects have two sets of wings for flying, gliding, and hovering. These fragile structures are strengthened by veins. In beetles, the front pair of wings has adapted to form hard wing cases. In flies, the two hind wings have become balancing organs called "halteres." An insect's six legs have strong muscles for walking, running, jumping, and swimming. Some flies can cling upside down because of the hairs and pads on the ends of their legs (see pages 112-113).

Vertical muscles contract, wings flap up.

Horizontal muscles contract, wings flap down.

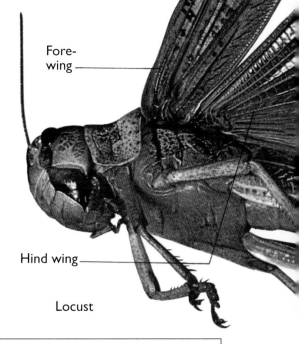

Fore-wing

Hind wing

Locust

Flapping the wings

Horizontal and vertical muscles are attached internally to the hard skeleton of the insect's thorax, which acts like a "click box." The sets of muscles contract alternately to make the insect's wings beat powerfully and rapidly. When the vertical muscles contract to pull down the roof of the thorax, it springs inward and the wings are flipped upward. When the horizontal muscles contract, the ends of the thorax are pulled in; the roof "clicks" back to its domed shape, and the wings are flapped downward.

Flight music

"The Flight of the Bumble Bee" is a piece of music by Rimsky-Korsakov. This Russian composer based his operas on fairy tales and folk-legends, and used the orchestra to make the sounds in the stories. The piece has a fast time, or tempo, which imitates the rapid buzzing of a bee's wings as it flies up and down looking for flowers.

Caterpillar tread

Many caterpillars have three pairs of proper legs at the front of their bodies, and five pairs of false legs, with suckers on the end, at the rear. The caterpillar moves one pair of legs at a time, distributing its weight equally over the other legs. This allows it to move over obstacles in its way. "Caterpillar tracks" based on the same principle are used on heavy vehicles like bulldozers, tractors, and army tanks. They distribute the vehicle's weight evenly so it can travel over rough or slippery ground.

Fleas jump using a "click box" mechanism similar to flying insects. They actually jump by relaxing their muscles, to allow the thorax to "click" outward. Fleas can jump up to 12 inches (30.5 cm), 130 times their own height!

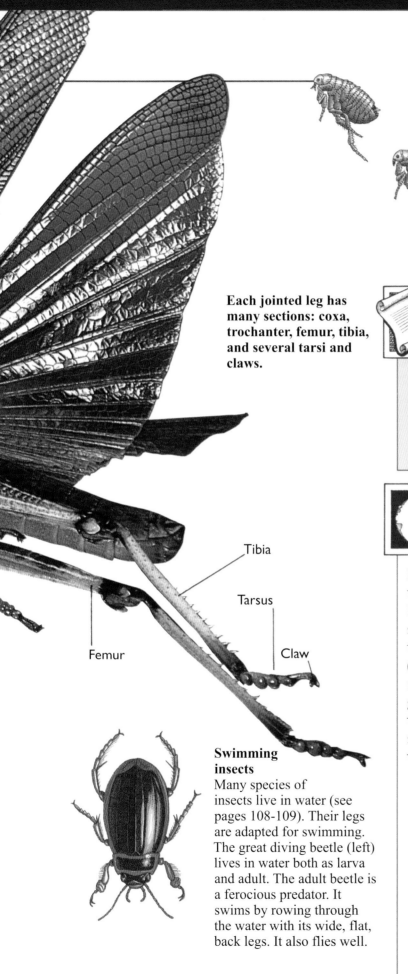

Each jointed leg has many sections: coxa, trochanter, femur, tibia, and several tarsi and claws.

Tibia

Tarsus

Femur

Claw

Performing fleas

Fleas are amazingly strong and have a good sense of balance. In Victorian times, these tiny animals were harnessed with fine wire, and made to walk tightropes in "flea circuses." The fleas were not "trained" to perform tricks, but were trying their best to escape.

Migrations

Many insects fly great distances to avoid winter cold. Monarch butterflies hold the long-distance record for such journeys, or migrations, as shown below. In September, these butterflies fly south from Canada, across North America, to Mexico. They travel up to 1,200 miles (1,931 km), 80 miles (129 km) per day. On reaching their destination, they hang in great clusters on pine trees and begin their hibernation, or winter sleep. In the spring, they drift back to lay their eggs on the Canadian milkweed plant.

Swimming insects
Many species of insects live in water (see pages 108-109). Their legs are adapted for swimming. The great diving beetle (left) lives in water both as larva and adult. The adult beetle is a ferocious predator. It swims by rowing through the water with its wide, flat, back legs. It also flies well.

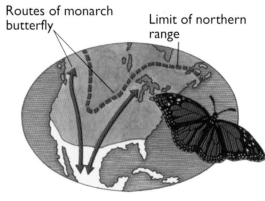

Routes of monarch butterfly

Limit of northern range

SENSES

In order to survive, insects must sense the world around them through sight, smell, hearing, taste, and touch. They are also sensitive to ultraviolet light, magnetism and gravity, temperature, and humidity. Many insect sense organs are extremely acute. In each species, they are tuned to the range of sensations most useful for survival. Messages from sense organs all over an insect's body pass along nerve fibers to the simple brain in its head. Insects cannot "think." They carry out predetermined responses to messages received from the sense organs, to feed, mate, lay eggs, attack, or escape.

Sensitive hairs
The heads, bodies, and legs of houseflies, like most insects, are covered by hairs, which are sensitive to movements in the air.

Compound eye
There are about 4,000 facets, or lenses, in a fly's eye. A bee has 5,000, a dragonfly has 30,000. Some ants have only nine.

Antennae

Taste organs are found on the feet.

Halteres
The fly's club-shaped halteres are modified hind wings, which vibrate to balance the insect, and measure its speed, and direction of flight.

Most insects have simple eyes, which perceive only light and dark shadows. They also have compound eyes, made up of hundreds or thousands of lenses, each seeing a slightly different view of the world. With these two different kinds of eyes, most insects can see all around them, in color, in fine detail, and even in the dark. The other main insect sense organs are the antennae. These projections are often covered with hairs attached to nerve fibers, which send messages to the brain whenever the hair is moved. There are also tiny hairs that are chemically sensitive to smells. The antennae also monitor moisture in the air.

Taste sensors are found on the mouthparts and often on the feet. Some insects have eardrums for hearing sounds. Others have hairs that are so sensitive, they can detect the air movements, or vibrations, made by sounds.

Through animal eyes
Insects see the world differently than other animals, such as birds and cats. Birds see in full color, with detail only in the center of the field of vision. Birds such as hawks can see small prey from a great distance (below). Cats focus well across a central, horizontal strip of their eye's retina. They see only dull colors. Each lens of an insect's compound eye sees the small scene in front of it. The brain merges the images to build up a detailed, three-dimensional picture.

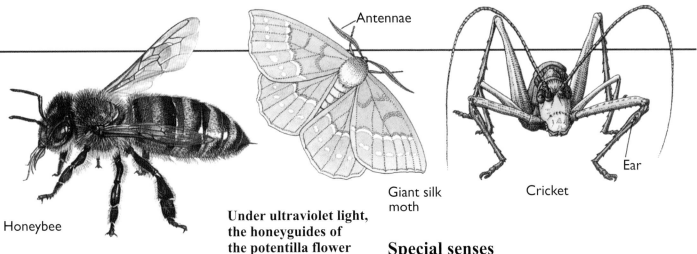

Honeybee

Antennae

Giant silk moth

Cricket

Ear

Under ultraviolet light, the honeyguides of the potentilla flower show clearly.

Special senses

Some kinds of insects have evolved special senses to help them thrive in their environment. A bee detects ultraviolet light, so it can follow lines on the petals of flowers, known as honeyguides, which lead it to the flower's nectar. Female giant silk moths produce tiny amounts of a special smell, called a "pheromone," to attract a mate. The male has long, feathery antennae which are so sensitive that he can locate the female from many miles away. Male crickets sing to attract females and to warn off other males by rubbing together special combs at the bases of their wings. Crickets have sensitive eardrums located on the forelegs to detect the song.

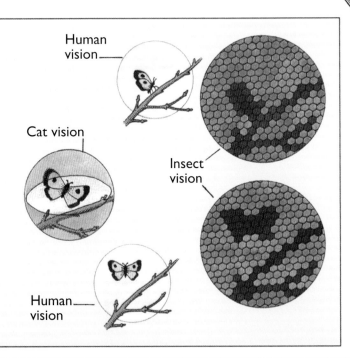

Human vision

Cat vision

Insect vision

Human vision

Literary crickets

In Victorian times, crickets often lived in houses. People thought disaster would befall the family if they stopped their song. Charles Dickens' story *The Cricket in the Hearth* describes a cricket merrily joining in with the kettle singing on the hearth.

Katydid crickets are named after the sound they produce. Two of these crickets "discussing" whether "Katy did" or "Katy didn't" prompted the author Susan M. Coolidge to tell the sorry story of *What Katy Did*. Jiminy Cricket is a character in the Walt Disney film of "Pinocchio," about a wooden puppet whose nose grew bigger every time he told a lie. The film is based on a story by Carlo Collodi.

POISONS AND STINGS

Animals use poisons for two reasons: to defend themselves against predators, and to overcome their own prey. Insects are no exception to this. Some have poisonous stings in their tails, and some bite with poisonous jaws. Others are just poisonous all over. Insects that use poisons to catch prey are often masters of disguise. Those that use poisons for defense usually advertise the fact by having bright warning colors on their bodies, typically red- or yellow-and-black. Other insects are not poisonous, but mimic those that are (see page 114).

Bee stings

Worker honeybees will defend the hive quite literally with their lives. The sting at the end of the worker bee's abdomen is a sac full of poison connected by a tube to a sharp, barbed spine. When the bee stings, the barbs make sure the sting stays in the victim while the venom is pumped into the wound. But that also means that when the bee flies off, the end of its abdomen is torn away and it then dies.

- Plates pump venom.
- Bulb full of venom
- Venom sac
- Worker bee

Sting remedies
Bees and wasps only sting when they feel threatened, so you are more likely to get stung if you shout or wave your arms to drive one away. If you are stung by a bee, remove the stinger with tweezers, taking care not to squeeze the poison sac. Wasps will not leave their stinger in your skin if you allow them to remove it.

Wash the wound thoroughly with antiseptic, and put a cold, damp cloth on it to relieve the pain. Bee and wasp stings are not dangerous unless the swelling blocks the throat, or unless the victim who has been stung has an allergy to insect stings.

Lethal weapon
The bodies of some kinds of insects are poisonous, and taste disgusting. This provides a good defense against predators who recognize the species, and do not attack it. Some squirt stinging liquids, while others have irritating hairs that get stuck in an attacker's skin. The grubs of a South African leaf beetle are so poisonous that Kalahari bushmen (right) use them to tip the ends of their arrows.

Ragwort is a poisonous weed, common in European fields. But the caterpillars of the cinnabar moth are able to feed on the plant, and store the poisons in their body tissues. A bird who eats one will become very sick. These caterpillars have yellow-and-black warning stripes on their bodies to advertise their identity. Birds learn after only one experience to leave them alone.

Insects in folklore medicine

The bodies of blister beetles contain an irritating fluid called "cantharidin," which these insects use to defend themselves against predators. Before modern medicines were developed, doctors used to apply this substance to their patients' skin as a treatment for warts. The blisters caused by the fluid were also thought to allow the escape of poisons that built up inside the body. Bee stings were thought to cure rheumatism, so bees were allowed to sting the inflamed joints of rheumatic patients.

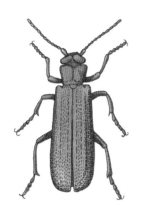

Proverbial insects

Traditional sayings or proverbs often refer to the familiar characteristics of common insects to help describe people's behavior. A group of people working very hard at a joint task are sometimes called "busy bees."
If someone has a particular concern, which others may not share, they are said to have a "bee in their bonnet." Children who will not sit still and concentrate at school are said to have "ants in their pants."
Can you describe someone who has a "butterfly mind?"

Bee in your bonnet

Beetle chemists

Bombardier beetles use a spectacular chemical reaction as a powerful weapon against attackers. The beetle has special chambers in its abdomen where it stores two chemicals, each fairly harmless on its own. When the beetle is alarmed, it mixes the chemicals in another chamber together with an enzyme, which aids the reaction. A rocket-like jet of hot, poisonous spray shoots from the end of the abdomen. The beetle can direct the spray by twisting its abdomen towards a victim. The boiling chemicals produced cause painful blisters.

BUGS

Bugs are a particular group of insects that share a common feature: they all pierce their plant or animal food, and suck the juices with mouthparts formed into a beak or long nose, called a "rostrum." The front pair of wings in many bugs is divided into two halves, a hard front part and a delicate, transparent back part. This gives the group its scientific name, "Hemiptera," or "half-wing." Cicadas, hoppers, aphids, and scale insects are all members of the family. Bugs undergo incomplete metamorphosis, and the young look very similar to adults. Many bugs are serious pests. Some, such as aphids, devastate plants; others carry disease, such as the assassin bugs of South America.

Many kinds of bugs are found all over the world, even at the edges of the sea. Different species are adapted to their particular environment. Bed bugs hide in crevices by day, and crawl into people's beds for a blood meal at night. Ponds teem with water bugs above and below the surface, as shown here. Shieldbugs feed on the sap of plants. European species are usually camouflaged—patterned to fit in with their surroundings. Tropical shieldbugs often have bright colors and patterns, and produce foul smells to ward off attackers. Young frog hoppers produce a nasty-looking foam known as "cuckoo-spit" as protection. Tree hoppers disguise themselves as thorns.

Many species of bugs live in or on the surface of fresh water. Water crickets and water measurers have water-repellent feet, which do not penetrate the surface of the water. They use their feet and antennae to sense the ripples caused by a drowning insect. Once a victim is located, it is stabbed with the insect's piercing mouthparts and its juices are sucked out. Different species of water bugs prey on tadpoles, beetle larvae, and other small creatures at different depths in the water.

Insects that live underwater must still breathe air. The water boatman solves this problem by trapping a layer of air in a bubble around its body.

Virgin birth

In the spring, aphid eggs hatch into wingless females. To save time, these insects do not mate or lay eggs, but give birth to live babies. Producing young without mating is called "parthenogenesis." It is quite common in insects. Later, winged males and females (right) are born. They fly off to mate and lay eggs for the following year.

Great diving beetles (page 103) store air beneath their wings, which can then be taken into the body through the spiracles (pages 98-99). The water scorpion (above) has a long siphon, like a snorkel, on the end of its tail, which it extends up above the surface of the water to breathe.

Opposite page:
This group of harlequin bugs from Australia consists of three red males, a yellow female, and two nymphs.

Song of the cicada
On warm summer evenings in tropical lands, in Mediterranean countries, and North America, the song of the male cicada is heard. The organ producing the sound is a "click box," like the wing mechanism (see page 102-103), located in the insect's abdomen. An area of hard cuticle is pulled in by a muscle and "clicks" out again. Long streams of clicks are produced at different pitches. They are amplified by air sacs in the abdomen. Cicadas have ears on their abdomens so that they can hear others singing. The males sing to attract a mate. Females respond by seeking out the best singer; other males respond by singing louder.

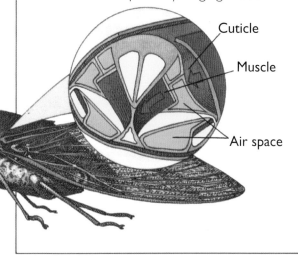

BEETLES

In terms of numbers, the group of beetles, or Coleoptera, has been more successful than any other kind of animal. There are at least 370,000 known species in the world, and new ones are being discovered all the time. Beetles are armor-plated insects. The head and thorax are covered in tough cuticle, formed into strange, threatening shapes in many species. Despite their heavy appearance, most beetles fly very well. Beetle grubs undergo complete metamorphosis to become adult.

Some species of beetles are herbivores (plant-eaters), and others are carnivores (meat-eaters). Some kill prey and eat it. Many perform the important function of consuming the dead bodies of animals, some eating the flesh, others eating fur or feathers. Some feed on animal dung. Some beetle pests consume grains or vegetables. Colorado beetles attack potato crops. Others attack vegetation, such as elm bark beetles that spread Dutch elm disease.

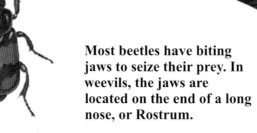

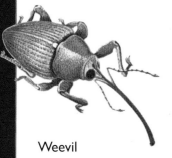

Weevil   Burying beetle

Most beetles have biting jaws to seize their prey. In weevils, the jaws are located on the end of a long nose, or Rostrum.

Light show

Glowworms, or fireflies, are neither worms nor flies. They are beetles that produce light to attract mates. During dark evenings, males and females flash signals to each other, like Morse code signals from a lantern. The code is different for each species. In Southeast Asia, whole trees pulse with thousands of these tiny lights. The light is made by a chemical reaction involving an enzyme that releases energy in the form of light.

Holy beetle
The female scarab beetle rolls a ball of dung to her burrow. She lays her eggs in the dung, and the larvae feed on it. The scarab beetle was sacred to the ancient Egyptians. They compared the insect's behavior with the action of their god Ra, who, they believed, rolled the sun across the sky each day. Egyptian craftsmen made scarab jewelry, using gold, lapis lazuli, and semi-precious stones.

Rove beetle

Chafer beetle

Insect machines
Some engineers have used insects as inspiration in the design and manufacture of machines. In the late 1940's, the vehicle manufacturer Volkswagen pioneered a family car with a rounded beetle shape. Its success was phenomenal, and over 19 million Volkswagen Beetles were produced and exported to nearly 150 countries worldwide.

Many kinds of beetles have fierce-looking jaws and horns. These are often for show, to frighten off predators, or for fighting between males. Stag beetles (left) are so named because the male has fearsome, antler-like jaws. Sparring stag beetles wrestle, each trying to turn his opponent over. In beetles, the front pair of wings form tough, often colorful wing cases called "elytra." These fold back when the insect is not flying, to protect the delicate wings beneath. In flight, the wing cases are raised.

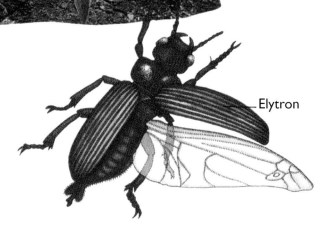
Elytron

Heralds of death
Deathwatch beetles are wood borers. The larvae live in the dead wood of trees or in cut timber, such as the roof timbers of a house. At mating time, the males and females call to each other from the tunnels they have bored, tapping their jaws on the wood, and making an ominous ticking noise. In the days before pest control and when illnesses were difficult to treat, this sound in old houses was thought to foretell a death in the family, ticking away the last minutes of someone's life.

FLIES

Flies are not a popular group of insects. Many are not beautiful, some have habits that humans find disgusting, and a few carry some of the world's worst diseases. But flies also recycle animal droppings and dead bodies, and pollinate many flowers. True flies, the Diptera, have two small, strong forewings. Their hind wings have become balancing organs, or halteres. Flies are a very versatile group, living almost everywhere in the world, even in the icy wastes of the Arctic.

Flies and disease

Blood-sucking flies transmit some of the world's worst tropical diseases. When they bite an infected person, they suck up tiny, disease-causing organisms, which they inject into another person at their next meal, and so the disease is passed on. Tsetse flies transmit sleeping sickness. Their victims experience tiredness and ultimately death. Assassin bugs carry Chagas disease in South America. Malaria, one of the world's most serious diseases, is spread by mosquitoes. Sandflies spread oriental sore or kala-azar, which destroys the skin and internal organs.

Hoverfly

Like all true flies, the cranefly has halteres, strong wings, and a large thorax. Hover flies can perform many skilled aerial maneuvers: they can spin round and fly backward as well as hover, as their name suggests.

Cranefly

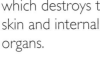
■ Tsetse fly
□ Assassin bug

■ Malaria mosquito
▨ Sandfly

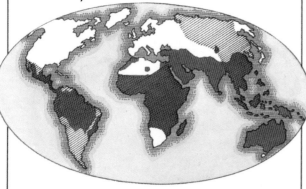

Flies are scavengers, feeding on almost everything: nectar, rubbish, blood, and flesh, dead or alive. They have a sucking proboscis, sometimes converted for piercing their prey. Fly grubs, often called "maggots" (top left), live in moist places, like stagnant mud or rotting meat. They undergo complete metamorphosis to reach their adult state. Flies are perfectly adapted for flying. They have a large brain and large eyes for extra control. Their halteres monitor speed, direction, and roll. The wings have special joints, which automatically twist the wing blades as they beat, to provide more lift. The enlarged thorax is packed with special flight muscles that contract rhythmically.

Walking on the ceiling

Houseflies and bluebottles have special sticky suction pads and hooks on the ends of their feet so they can crawl up windows or upside down across the ceiling. The hairs all over their body are so sensitive to air movements that they can feel danger, such as a flyswatter, coming just in time to get away.

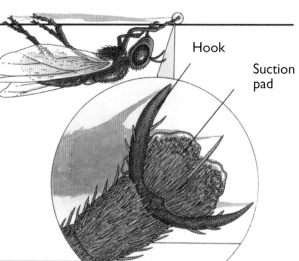

Hook

Suction pad

Hygiene

It is important to keep food covered if flies are around. House flies feed on rubbish and dung. Their feet and mouth-parts become contaminated with germs, which can be transferred onto food and kitchen surfaces if they alight there.

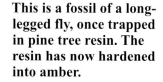

This is a fossil of a long-legged fly, once trapped in pine tree resin. The resin has now hardened into amber.

Fly-eating plants

The Venus fly trap plant lives in the swamps of North America, where the soil is poor. The plant catches flies to supplement its diet with extra nutrients. The fly trap's leaves have sticky, sensitive hairs and toothed edges. When an insect settles to feed, it disturbs the hairs. This triggers the two halves of the leaf to snap together, trapping the insect.

Metamorphosis

The theme of changing from one form to another has intrigued writers and artists for many centuries. The Roman poet, Ovid, wrote a long poem called *Metamorphoses* **over** 2,000 years ago. Based on mythical tales, it tells of characters, such as Daphne, who changed into a laurel tree to avoid the attentions of the god Apollo, and Narcissus, who changed into a flower. Artists such as the Spanish painter Salvador Dalí have been inspired by the same theme.

The idea of humans changing into insects has also held horror and fascination. Czech writer Franz Kafka wrote a strange story about a man who wakes up one morning having mysteriously changed into a huge insect overnight. The film called *The Fly* is a horror story about a scientist who slowly becomes a fly.

The fly trap's leaves give off a liquid containing digestive chemicals called "enzymes," and the insect's body is slowly dissolved and absorbed.

113

BUTTERFLIES AND MOTHS

Butterflies and moths are called "Lepidoptera," which means "scale-wings." Their wings are covered with tiny scales, arranged like rooftiles. Some scales have beautiful colors, and others bend light, like crystals, to give a rainbow sheen. Butterflies are generally more colorful than moths. They fly in the daytime, and hold their wings together upright when resting. Their antennae are club-shaped. Moths are active at night. They hold their dull-colored wings flat over their backs when resting. Some male moths have feathery antennae. Butterflies and moths undergo complete metamorphosis.

Owl butterfly (*Caligo oileus*)

Butterfly mimics

The colorful markings on some butterflies' wings are warning colors, to deter predators. Poisonous insects advertise their distastefulness in this way. Birds soon learn to recognize these species and avoid them. Different species of poisonous moths or butterflies from the same region reinforce the message, by having very similar patterns and wing shapes. In Peru, two poisonous species, heliconius and podotricha butterflies, look alike. This is called "Mullerian mimicry" (imitation). Some other butterflies that are not poisonous "cheat" by mimicking the warning patterns of poisonous species. In North America, a species of harmless viceroy butterfly looks very like the poisonous monarch butterfly. This is called "Batesian mimicry."

Podotricha telesiphe

Heliconius telesiphe

Siderone galanthis

Viceroy

Monarch

Butterflies and moths, like all insects, are cold-blooded. Their body temperature is about as warm or cold as their surroundings, since they cannot generate their own body heat as warm-blooded mammals and birds can. Butterflies spread their wings in the sunshine to warm up and hide in the shade when they are too hot. Moths have furry bodies to retain the heat they absorb during the day so that they can fly at night.

Disgusting disguises

The young of many moths and butterflies camouflage or disguise themselves as inedible objects to avoid being eaten by predators. Geometer moth caterpillars look like twigs, and position themselves on branches so as to complete their disguise. The European black hairstreak chrysalis and the hawkmoth caterpillar from Central America (right) pretend to be unpleasant bird droppings.

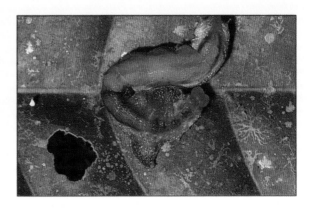

Large elephant hawkmoth

Morpho butterfly (*Morpho menelaus*)

Dasyopthalma rusina

Adapting to industry

The peppered moth is an example of how, over many generations, some species of insects are able to adapt their camouflage to fit in with a changing environment. The normal form of the peppered moth is creamy with dark speckles, difficult to see on the bark of trees. But following the Industrial Revolution in the 19th century, a new form of dark moth became common, which could hide on sooty bark.

Insect painter

Jean Henri Fabré (1823-1915) lived in Provence in France. He was a village school teacher before he turned to entomology, the study of insects. But he did not collect dead insects, like many naturalists of the time. He studied their habits by watching them in the wild.

Fabré wrote many books on insect behavior, describing each detail of their lives. He also left behind many beautiful watercolor paintings of the species he had studied.

Madame Butterfly

The opera *Madame Butterfly* was written by Italian composer Giacomo Puccini in 1904. It tells of a tragic love affair between an American naval officer, Pinkerton, and a Japanese girl, Butterfly. They marry, but Pinkerton leaves, and returns years later with a new wife. Puccini's beautiful melodies convey the drama, passion, and tragedy of the story.

BEES AND WASPS

Bees, wasps, and ants all belong to the same group of insects, called "Hymenoptera," meaning "transparent wing." Bees and wasps have two pairs of wings, and a narrow "waist." Some have yellow and black warning colors, and a poisonous sting. The life cycle involves complete metamorphosis. Many species are solitary; honeybees, bumblebees, and common wasps create colonies where the eggs and grubs are cared for by family members.

All bees and some wasps make nests for the young. The sand digger wasp digs holes for its eggs, and the mason bee tunnels in cement. The leaf-cutter bee constructs leaf-lined chambers. Honeybees live in nests, either in holes in trees or in man-made hives. The queen mates with the male drones in flight, and lays her eggs, most of which develop into sterile female workers. Worker honeybees make and repair the nest, and go out to collect pollen and nectar to make into honey, to feed new grubs, and sustain the bees in winter.

Construction work
Honeybees mold perfect hexagonal cells from beeswax, which comes from a gland below their abdomen.

Paper wasps fashion their delicate nests from chewed wood fibers, below.

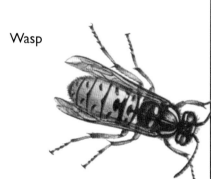
Wasp

Dancing bees
When a scouting honeybee returns to the hive laden with pollen and nectar, the other worker bees gather to find out the location of the new food. The bee performs a round dance on the vertical surface of the honeycomb. This dance becomes a figure eight, the bee waggling its rear excitedly as it passes across the center of the figure. The angle of the waggle to the vertical is the same as the angle between the sun, the hive, and the food. The amount of waggling indicates how distant the food source is.

Round dance

The scout stops during the round dance to give the other bees samples of pollen and nectar.

Figure eight

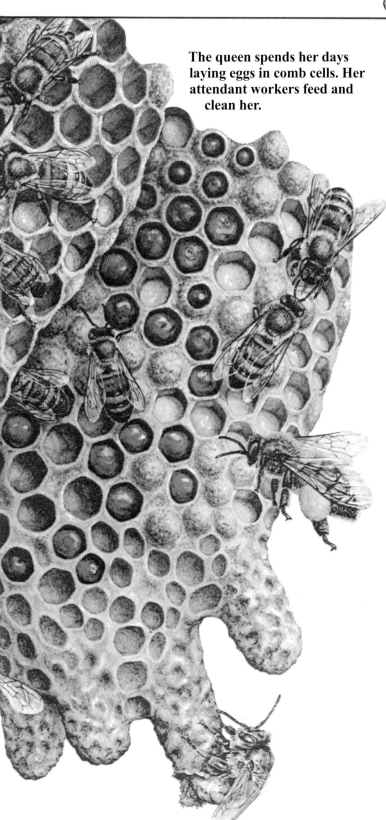

The queen spends her days laying eggs in comb cells. Her attendant workers feed and clean her.

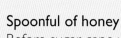

Spoonful of honey
Before sugar cane was brought to Europe around 700 AD, people used honey to sweeten their food. Egyptian tomb paintings show honey and beeswax being harvested from man-made hives.

Pollination
For a flower's seeds to develop, it must be fertilized by pollen from the same or another flower. The pollen can be dispersed by the wind, or transferred on the bodies of insects, such as bees. The insects are attracted by the colors, perfume, and sweet nectar of flowers. As they wander over the petals, their hairy bodies become dusted with powdery pollen. When they visit another flower, the pollen dusts off and pollinates it. The pollinating activities of bees are even more important to the farmer than their honey-making duties.

Waspish fashion
Dresses with elegant "wasp" waists were created in the 19th century by French fashion designer Charles Frederick Worth. The wasp waist was created by a tight corset. Unfortunately, the fashion was extremely uncomfortable, making some women faint.

Pollen baskets
Honeybees have hair and notches on their legs. They use them to comb flower pollen from their heads and bodies, and pack it into pollen baskets, fringed with long, stiff hairs, on each back leg.

ANTS AND TERMITES

Ants belong to the group of Hymenoptera, like bees and wasps. Termites belong to the order of Isoptera, meaning "equal wing." Nevertheless, ants and termites have very similar lifestyles. They are mainly social insects, living in huge families or colonies, where each insect has a particular job to do. Most do not reproduce; their lives are devoted to caring for their sisters and brothers. Only the queen mates and lays eggs. Her many young (the workers) build, repair, and defend the nest.

Caste of thousands
Different kinds, or castes, of ants or termites perform different jobs in a colony. Worker ants tend to the queen (left), and the grubs, and pupae (right). Others clean the nest (above right) and go out in search of food. Soldiers defend the colony.

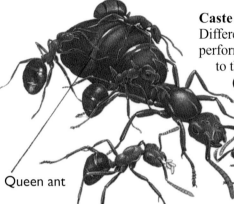

Queen ant

Workers

Pupae

Social ants
Most ants have poor eyesight, but a good sense of smell. They communicate with nest members through touch and through scents called "pheromones," which they produce. When foraging ants find food, they lay a scent trail for others to follow. Worker ants produce a different scent if they find a damaged part of the nest, which brings others to help with the repair. Ants from one colony recognize each other by their smell, and will attack an intruder from a different colony. They defend themselves by biting and squirting stinging formic acid into the wound they have made. Ants feed on many different types of food. A column of army ants will tear apart and carry off any small creature in its path. Each ant can lift a load many times its own weight.

Making an ant home
You can study ants more easily by building an ant home from a glass tank or plastic box. Cover the outside of the tank with dark paper. Half-fill the tank with earth, and stock it with small black or red ants from the garden. Add damp soil and leaves. After a few days, remove the paper to see the tunnels built against the sides.

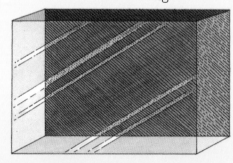

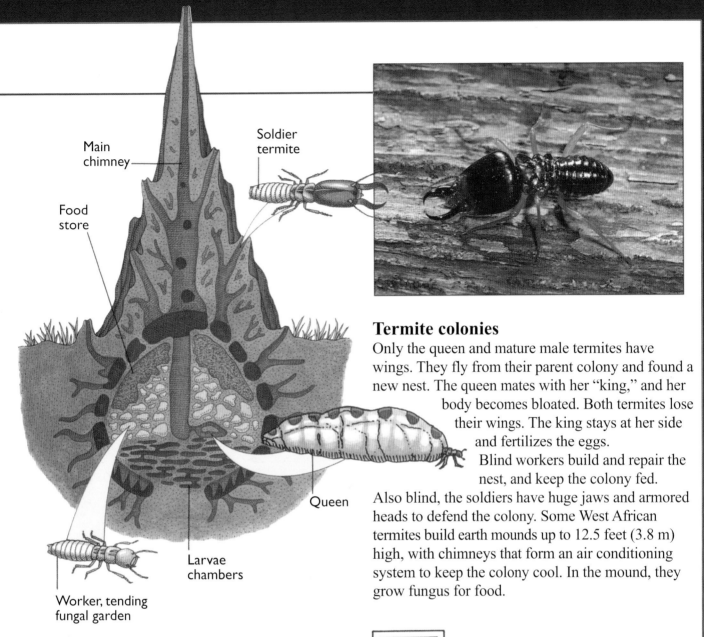

Main chimney
Soldier termite
Food store
Queen
Larvae chambers
Worker, tending fungal garden

Termite colonies

Only the queen and mature male termites have wings. They fly from their parent colony and found a new nest. The queen mates with her "king," and her body becomes bloated. Both termites lose their wings. The king stays at her side and fertilizes the eggs.

Blind workers build and repair the nest, and keep the colony fed. Also blind, the soldiers have huge jaws and armored heads to defend the colony. Some West African termites build earth mounds up to 12.5 feet (3.8 m) high, with chimneys that form an air conditioning system to keep the colony cool. In the mound, they grow fungus for food.

Feed ants with ripe fruit, meat, or jam, and provide fresh leaves and water on a damp paper towels. Keep your tank in a cool place, and cover it when you are not studying ant behavior, so air can get in, but the ants can't escape. If you have managed to catch a large queen with your stock, the colony should go on indefinitely, and may even produce a swarm of winged ants in the summer. Let them go to produce a new nest.

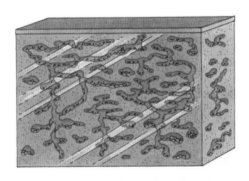

Anteaters of the world

Ants and termites are food for any animal with strong claws to rip open the nests, and a long, sticky tongue to lick the insects out. In Central and South America, armadillos and giant anteaters (below) live on the savanna, and collared anteaters are found in the forests. The aardvark lives off the same diet in South Africa, and the pangolin in Asia. The spiny anteater lives in Australia.

119

INSECTS AND PEOPLE

Humans have lived with insects since we first evolved. Our ancestors must have battled against biting fleas and scavenging cockroaches, just as we do today. Many insects are pests; some spread disease, some damage property, others consume crops. But insects also play a very positive role. They provide a vital food source for animals such as birds and reptiles. Some pollinate flowers, and some dispose of waste matter. Insects provide silk and honey, and delight our eyes and ears. The world would not be the same without them. But as the world's wild places are disappearing, so are many insect species.

We call insects "pests" when they become so numerous that they begin to threaten human health and comfort. There are thousands of insect pests. They range from the irritating houseflies and head lice, which can infest our homes and bodies, causing discomfort, to mosquitoes and locusts, which can cause death and widespread devastation. Many insects only become pests because humans choose to cultivate the food they eat on a large scale, providing hundreds of acres, or tons, of their usual diet.

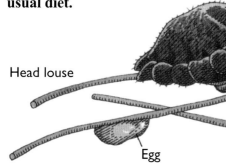

Head louse

Egg

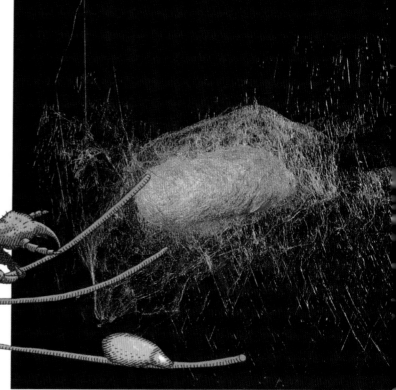

Agricultural pests

There are many serious insect crop pests around the world. Some eat the leaves or roots of plants. Others infect the plant with virus disease. Both result in crop failure, with financial losses in richer nations, and starvation in developing countries. Rice and maize (sweet corn) are two of the most important crops in the world. The brown plant hopper damages rice harvests, while the corn leaf hopper is a serious pest of maize crops. The Colorado beetle devastated North American potato crops in the 19th century, until the use of pesticides finally brought it under control.

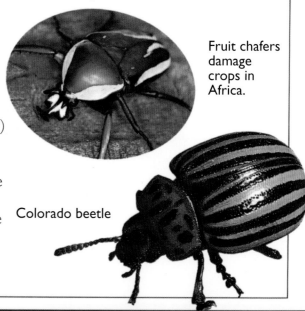

Fruit chafers damage crops in Africa.

Colorado beetle

CLASSIFICATION OF INSECTS

Scientists classify insects into groups that share similar characteristics, such as appearance and behavior. All insects belong to the animal kingdom under the class Insecta. This class is divided into 31 orders, which appear below. The names of the orders are indicated in bold type. The orders are subdivided into genera, and the genera into closely related species, or individual kinds of insects. The chart below shows a typical insect from each order, identified in normal type.

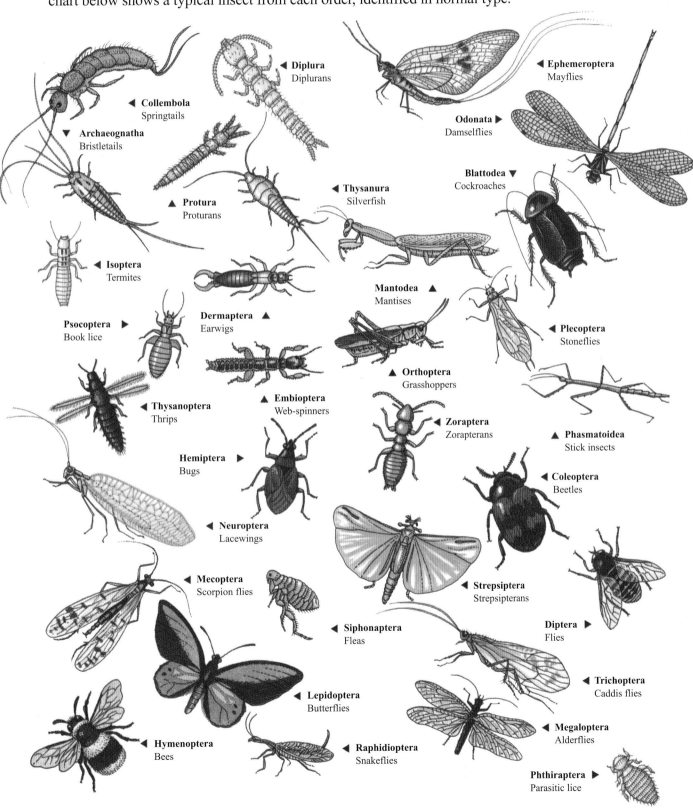

Photo credits
The majority of the pictures are from Bruce Coleman Ltd apart from pages:
Front cover bottom, Roger Vlitos,
Robert Harding Picture Library, Hulton Deutsch, Science Photo Library, Planet Earth Pictures, H. Leverton Ltd, Frank Spooner Pictures, Oxford Scientific Films, Spectrum Colour Library.

I KNOW ABOUT! DINOSAURS

SOFT TISSUE AND DNA
128

DINOSAUR TRACKS 130

NESTING DINOS 132

T. REX REVEALED 134

HORNED DINOS 140

BIRDS THAT FILL THE GAP 144

EXTINCTION THEORIES 146

Dinosaur World

Dinosaurs were the dominant land animals throughout the Jurassic and Cretaceous Periods (206-65 million years ago), but their world was very different from ours. Landmasses lay in different positions from today and formed vast supercontinents, such as Gondwanaland and Laurasia (below). For long periods of time, the continents were connected and dinosaurs could walk to all parts. Early in the Jurassic and again in the Cretaceous, sea levels rose and areas of land were flooded by seas, isolating dinosaurs on smaller island continents.

The world was warmer than today and without extreme changes in the seasons. Mild conditions spread to the Poles and there were no ice caps.

THE WORLD OF THE DINOSAUR

Conditions during the time of the dinosaurs were a lot hotter and wetter than they are today. This warm and often humid atmosphere gave rise to lush, dense vegetation (above). At ground level, there were ferns, and in wetter areas, horsetails. Woods and forests were dominated by palm-like cycads, gingkoes, and conifers.

FLOWERS
Plant communities changed dramatically in the Cretaceous with the appearance of flowering plants (left). These very successful plants soon dominated plant life.

Dinosaurs shared the plains, forests, and swamps with many other groups of animals. Pterosaurs and primitive birds flew overhead, while lizards scampered under their feet (right). Turtles, frogs, and crocodiles inhabited lakes, rivers, and marshes and our ancestors, the early mammals, hid in trees or among rocks.

SEA LIFE

During the time of the dinosaurs, the seas were dominated by marine reptiles. Ichthyosaurs, the fish lizards, were very similar in size and shape to dolphins and probably fed on fish and squid.

Plesiosaurs, the top predators (right), had powerful jaws and rows of sharp, pointed teeth, which they used to kill and tear apart their prey. They were up to 39 feet (almost 12 m) long and had a small head on top of a long, snaking neck.

DRAGONS OF THE AIR

The skies above the dinosaurs' heads were filled with pterosaurs (left). With broad, membranous wings and fur-covered bodies, most pterosaurs fed on fish or insects that they caught while flying. Early pterosaurs were about the size of a crow. However, later forms grew very large, reaching giant size with wingspans of 39 feet (almost 12 m), the size of a small airplane.

EARLY MAMMALS

The first mammals evolved from a group of reptiles that existed long before the arrival of the dinosaurs. Early mammals, such as Megazostrodon (right), were hairy and usually no bigger than a hamster. They did their best to avoid predatory dinosaurs by only coming out at night. With their sharp eyesight and sense of smell, they caught and ate insects, snails, slugs, and other small delicacies.

Soft tissue and DNA

Usually only the toughest parts of dinosaurs, their bones and teeth, survive to become fossils. Soft tissues, such as the internal organs, muscles, and skin, decay very quickly or are eaten by predators or scavengers. But sometimes dinosaurs were killed and buried very quickly—some were smothered by huge sandstorms, while others fell into dark, poisonous sediments at the bottom of lakes—and parts of their soft tissues were preserved. Skin is the most likely to be found, because it is relatively tough, but remains of muscles have recently been discovered, and some scientists claim to have found blood cells and even fragments of dinosaur DNA.

Real DNA from dinosaurs?

U.S. researchers claim to have extracted DNA from well-preserved dinosaur bones, and Chinese scientists claim to have recovered it from a dinosaur egg. New tests show, however, that it is modern, not ancient, DNA and probably comes from contamination of the specimens when they were handled by humans. In the film *Jurassic Park* (left), scientists brought dinosaurs back to life by growing them from cells into which they had injected dinosaur DNA, extracted from dinosaur blood found in the gut of blood-sucking flies preserved in amber. Theoretically, this technique is possible, but there is less than a one-in-a-billion chance that a complete DNA sequence could be found and extracted. The cost of such a technique would be enormous.

Because soft tissues affect the skeleton, paleontologists can use the shapes of bones to try to reconstruct the muscles, guts, and brains. Knobs and crests on limb bones tell us much about the size and position of muscles, internal casts of the brain case reveal the shape of the brain, and the size and shape of the ribs shows how big the guts were.

Girder-like spine supports neck

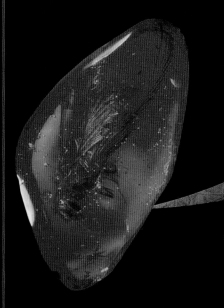

Insect preserved in amber

Entombment in amber is one of the best ways of becoming a fossil. Bones, soft tissues, and stomach contents are preserved in the finest detail. No dinosaurs have ever been found in amber, but lizards, mammals, and insects (left) have been preserved this way.

If most or all of the skeleton is found, paleontologists are able to reproduce how a dinosaur looked and behaved with a high degree of accuracy. Once the muscles have been placed over the skeleton, the whole animal can be covered in its skin (right). Although some skin remains have been found, they only tell paleontologist the skin's texture. For an idea of its color, scientists look to animals that are alive today and have a similar way of life to the dinosaur, (i.e. if they hunters or grazers.)

Reconstruction of dinosaur muscle

WHAT IS DNA?

DNA (deoxyribonucleic acid) is the recipe that cells use to build living organisms. The DNA is stored in the nucleus, which acts as the command center of the cell. Under rare conditions, bits of DNA have been preserved in fossils.

Pelecanimimus

Traces of soft tissue around the skull of Pelecanimimus (right), an unusual new ornithomimosaur found in Spain, show that this small dinosaur seems to have had a small crest on the back of its head and a pouch in the throat region. Perhaps Pelecanimimus was a fish-eating dinosaur, which grabbed prey using its 200 or so small, sharp teeth, and stored them in the throat sac, much like a pelican does today.

Large spikes
Studded skin
Tail club

Scaly hide

Impressions of scaly skin (left), much like that of living reptiles, have been found in the fossils of two plant-eating hadrosaurs that died in dry semi-desert conditions, and became mummified. The armored skin of four-legged ankylosaurs (above) had bony plates and spikes. They also had clubs on the end of their tails, which they could swing violently to defend themselves.

Dinosaur Tracks

In recent years, there has been a great upsurge of interest in dinosaur tracks. These tracks are very important because they provide the only direct evidence of how dinosaurs lived. From tracks we can tell how dinosaurs stood, how they moved, how fast they could run, and whether they lived in herds or on their own.

Many tracks show that some dinosaurs, such as sauropods and ornithopods, lived in herds. Lots of tracks heading in the same direction, equally spaced out and all bending at the same time, show that they were made by a herd passing by, and not by lots of single animals over a long period of time.

Dinosaur tracks

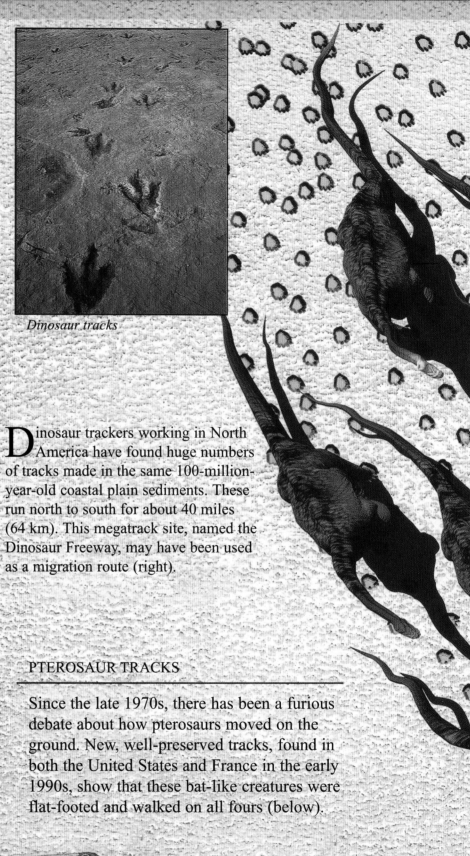

Dinosaur trackers working in North America have found huge numbers of tracks made in the same 100-million-year-old coastal plain sediments. These run north to south for about 40 miles (64 km). This megatrack site, named the Dinosaur Freeway, may have been used as a migration route (right).

PTEROSAUR TRACKS

Since the late 1970s, there has been a furious debate about how pterosaurs moved on the ground. New, well-preserved tracks, found in both the United States and France in the early 1990s, show that these bat-like creatures were flat-footed and walked on all fours (below).

Dinosaur footprint

DEATH BY DINOSAUR

Life was dangerous in the time of the dinosaurs. Even if you were not caught and eaten, you could be crushed to death. This is what happened to a group of shellfish, trampled to death by a sauropod paddling along in a shallow lake in Colorado 150 million years ago.

MARTIN LOCKLEY

Discoverer of many important dinosaur tracks and author of numerous articles and books on dinosaur tracking, Martin Lockley, a British paleontologist who lives and works in Colorado, has rescued paleoichnology (the study of fossil tracks) from its backwater status and propelled it into the mainstream of paleontological research. In his career, Lockley has described many important new finds, including the prints of Tyrannosaurus, the tracks of pterosaurs, and the Dinosaur Freeway (see left).

STAND UP STRAIGHT

Paleontologists have long argued whether ceratopsians like Triceratops had upright forelimbs, or whether they stuck out to the side, like lizards. Tracks made by Triceratops and recently found in rocks near Denver, Colorado, show that ceratopsians stood upright when they walked (left).

Tracks have shown some unusual features of dinosaur behavior. In one example (above), one of the toes from the left foot of a therapod was missing. The tracks also show that the dinosaur walked with a limp, as shown by the short, irregular steps it made.

Nesting Dinos

Paleontologists have long known that dinosaurs laid their eggs in nests, and thought that, like other reptiles, they left their young to hatch out on their own and fend for themselves. But, an astonishing discovery made in the Gobi desert in 1993 shows that some may have sat on the nest. A complete skeleton of Oviraptor, with its legs tucked beneath it and arms curled around its sides had been buried in a sandstorm while sitting on a nest of 22 eggs.

Egg of therizinosaur

NESTING GROUNDS

Dinosaur nesting grounds containing lots of closely spaced nests that seem to have been used year after year were first reported in the 1980s. New discoveries in Mongolia show that sometimes different kinds of dinosaurs nested together. Indian paleontologists have just discovered a dinosaur hatchery that contains the fossilized eggs of titanosaurs (large sauropods) and might be 600 miles (966 km) long.

Clutches of dinosaur eggs were laid in many different ways: Most were laid in circles (above), but some were laid in rows, either in arcs or in straight lines.

The discovery of an Oviraptor sitting on a nest of eggs (right) has altered paleontologists' view of the role of dinosaur parents. They now appear to have cared for their young until they were quite developed.

Huge numbers of dinosaur eggs have recently been found in cretaceous rocks in China (right). Among these are the largest eggs ever found and the eggs of therizinosaurs containing complete and beautifully preserved embryos at various stages of development. It is very rare for dinosaur embryos to stay this intact.

Chinese dinosaur eggs

EMBRYOS

Embryos are very rare, but sometimes, as in the case of this Oviraptor embryo (left and below) discovered in Mongolia, they are found still in the egg. Not only does this tell us what dinosaur embryos looked like, it also tells us which kind of dinosaur laid the egg!

EGG THIEF

Oviraptor (above) was long thought to be an egg-thief that gobbled up the eggs of Protoceratops, a small sheep-sized ceratopsian (see page 140). It was believed that Oviraptor would crush eggs between its large, curved, horny jaws. In fact, it was a dutiful parent staying with its unborn young even to the point of death.

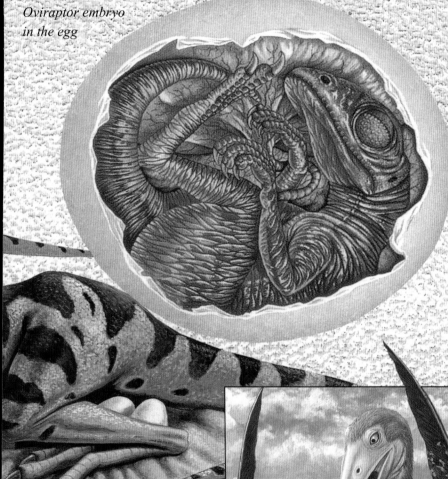

Oviraptor embryo in the egg

Like modern-day birds, baby dinosaurs may well have used egg teeth to chip their way out of the egg (above). After a while, this egg tooth may have fallen off. The young dinosaur may also have been helped out of the egg by the parent. Alternatively, the egg may not have had a tough shell, but, like the egg of a turtle, may have been quite leathery and relatively easy to hatch out of.

Tiny, fragile bones of baby pterosaurs were recently found by Russian paleontologists in sands laid down in an estuary in central Asia some 80 million years ago. The babies, which must have been looked after by their parents (right), probably fell from nests built in large plantain trees bordering the estuary.

Mother and baby pterosaur

T. rex REVEALED

For many, Tyrannosaurus rex is symbolic of dinosaurs as big, fierce, and extinct. But we know surprisingly little about T. rex. It is only in the last few years that complete skeletons with skulls have been found. Much work remains to be done on these before T. rex gives up all its secrets.

For many years, T. rex was thought to have stood and walked in an upright position, but new models based on studies of well-preserved fossils show a much more aggressive, agile-looking animal with its head thrusting forward and a long, stiff tail counterbalancing it behind.

At 40 feet (over 12 m) long and a weight of at least 6 tons, T. rex was one of the largest land predators of all time.

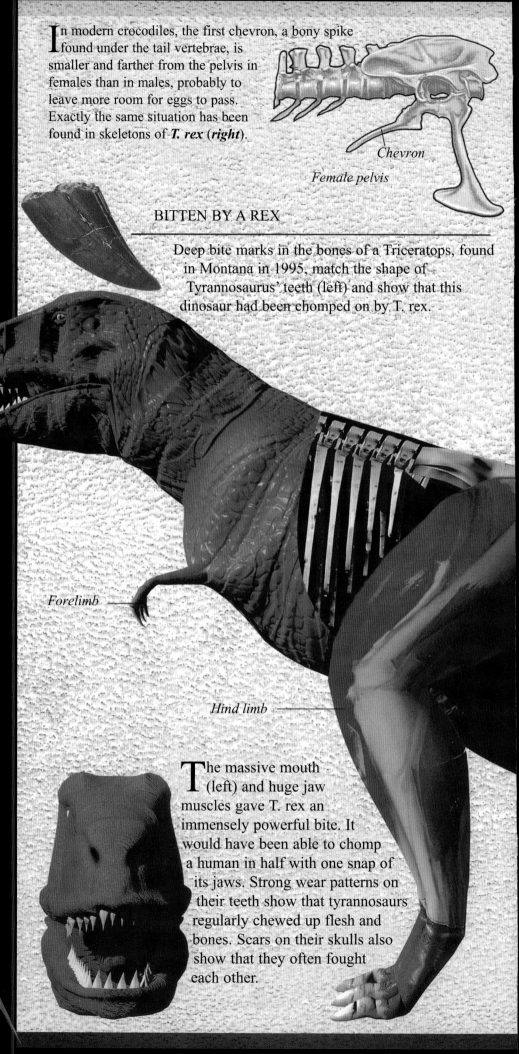

In modern crocodiles, the first chevron, a bony spike found under the tail vertebrae, is smaller and farther from the pelvis in females than in males, probably to leave more room for eggs to pass. Exactly the same situation has been found in skeletons of **T. rex** (**right**).

Chevron

Female pelvis

BITTEN BY A REX

Deep bite marks in the bones of a Triceratops, found in Montana in 1995, match the shape of Tyrannosaurus' teeth (**left**) and show that this dinosaur had been chomped on by T. rex.

Forelimb

Hind limb

The massive mouth (**left**) and huge jaw muscles gave T. rex an immensely powerful bite. It would have been able to chomp a human in half with one snap of its jaws. Strong wear patterns on their teeth show that tyrannosaurs regularly chewed up flesh and bones. Scars on their skulls also show that they often fought each other.

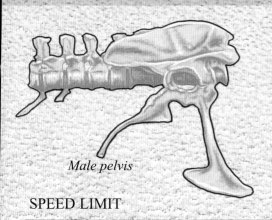
Male pelvis

Albertosaurus

Close relatives of T. rex include Albertosaurus, Daspletosaurus, and Tarbosaurus. All lived at about the same time as Tyrannosaurus, in the same region, and were about the same size and shape. A new relative, Siamotyrannus (below), was found in Thailand in 1993. A lot smaller than Tyrannosaurus and at least 30 million years older, it gives us some idea of what T. rex's ancestors looked like.

SPEED LIMIT

New studies on the design of the skeleton of Tyrannosaurus have found that this dinosaur (below) was so heavy that it would be badly hurt or possibly killed if it fell over while running at more than 20 mph (32 km/h). But Tyrannosaurus rex didn't need to run fast—there was nothing to run from and prey could always be ambushed and killed with a single bite.

Siamotyrannus pelvis

Counterbalancing tail

A new study by the American paleontologist Tom Holtz confirms ideas that T. rex is more closely related to coelurosaurs (small, lightly built two-legged hunters) than any other dinosaurs. A vital clue comes from the shape of the foot bones (left). In both the feet of Tyrannosaurus and the coelurosaurs, the middle bone of the three toes is pinched slightly at the top (left). This shows that Tyrannosaurus was a super-heavyweight coelurosaur.

Pinched foot bone

Foot of T. rex

WHO OWNS SUE?

Sue, discovered in South Dakota in 1990 by Susan Hendrickson, is one of the most complete and best preserved examples of T. rex (below). Unfortunately, there is a dispute over who owns Sue: the people who found her, the man who sold it to them or, the Sioux Indians who own the land. Until this is sorted out, Sue is being held "under arrest" by the FBI in Rapid City, South Dakota.

Theropod Predators

Two-legged theropods, including T. rex, were some of the largest and most ferocious predators of the dinosaur age. Newly discovered remains of one, Carcharodontosaurus, including an almost complete skull, reveal the huge size of this African dinosaur. Found in 1995 by an American expedition to the Atlas Mountains in Africa, this dinosaur had a skull that was 5 feet (1.5 m) long (below) and may have been even larger than T. rex. Carcharodontosaurus was a frightful predator, capable of killing and consuming other large animals. It shared its habitat with Deltadromeus, a smaller predator (right).

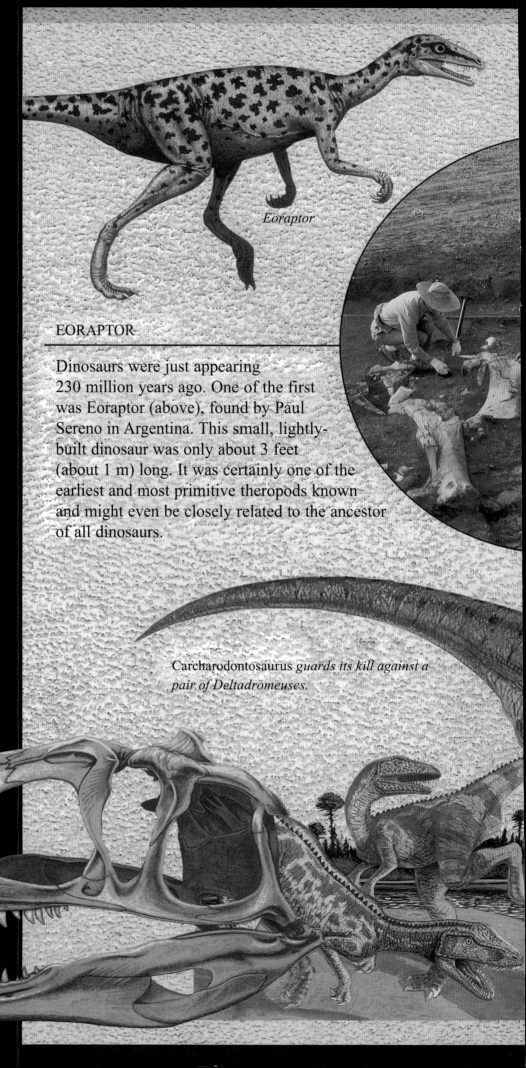

Eoraptor

EORAPTOR

Dinosaurs were just appearing 230 million years ago. One of the first was Eoraptor (above), found by Paul Sereno in Argentina. This small, lightly-built dinosaur was only about 3 feet (about 1 m) long. It was certainly one of the earliest and most primitive theropods known and might even be closely related to the ancestor of all dinosaurs.

Carcharodontosaurus *guards its kill against a pair of Deltadromeuses.*

Carcharodontosaurus skull

Giganotosaurus

The theropod Giganotosaurus (left) lived about 100 million years ago and may have been the largest land predator ever.

The skull is about the same size as that of Tyrannosaurus, but its leg bones are a little longer and sturdier, suggesting that this animal was heavier and even more massive than Tyrannosaurus.

AFROVENATOR

In 1993, an expedition to Niger, Africa (left), led to the discovery of Afrovenator (below), a large predatory dinosaur that was 30 feet (about 9 m) long.

Afrovenator

Between 1986 and 1990, joint Chinese-Canadian expeditions to northern and western China found large numbers of new fossils including Sinraptor, Monolophosaurus, and Sinornithoides. Many of these new specimens are now on view in Chinese museums (above).

BARYONYX

Finding a skeleton is only the first step in a long task of collecting a new dinosaur. It took more than ten years for the staff at the Natural History Museum in London, England, to free the bones of Baryonyx, a predator, from the rock in which it was buried 120 million years ago (below).

Alxasaurus Mystery

Ever since the 1950s, paleontologists in central Asia have been finding odd pieces of some very peculiar, large, long-armed dinosaurs from the Cretaceous era. The first bones to be found were thought to be those of turtles, and even when they were identified as a dinosaur, no one could agree as to what kind of dinosaur it was.

The discovery of a nearly complete skeleton of Alxasaurus (main picture) has cleared away much of the mystery. Alxasaurus is a therizinosaur and shows that lots of other odd dinosaurs belong to this group. But what did therizinosaurs use their peculiar teeth, long arms, and huge claws for?

Alxasaurus eating ginkgo leaves next to a Psittacosaurus.

LIFESTYLE

The odd shape of Alxasaurus' teeth (see right) suggests that it fed on plants, such as the leaves from ginkgo trees (below). Sitting or squatting on its haunches, Alxasaurus dragged branches and fronds toward its mouth using its powerful arms and long claws (bottom left). Alternatively, by rearing up on its hind limbs and supporting itself with its forelimbs, Alxasaurus could reach up with its long flexible neck to strip leaves from the tops of small trees.

Gingkoes (left) are one of the few remaining examples of non-flowering plant groups known as gymnosperms. The Ginkgo tree was once abundant in the time of the dinosaurs but is now only found wild in China.

HABITAT

Fossils collected from the sediments in which the bones of Alxasaurus were preserved show that this animal lived in a well-vegetated river valley, which included conifers, ferns, and flowering plants. Alxasaurus shared this environment with turtles, the crocodile-like champsosaurs, and Psittacosaurus, a small, bipedal, plant-eating dinosaur (below).

WHOSE RELATIVES?

Therizinosaurs have such a peculiar body that it has been very difficult to find out which major group of dinosaurs they might belong to. Some paleontologists put them in a separate group all of their own, some thought that they might be related to sauropods, while others put them in the middle between sauropods and theropods. However, newly discovered remains show that they are theropods.

SKULLS AND TEETH

The preserved skull of Erlikosaurus is the best evidence we have to tell us what the head of therizinosaurs looked like. The skull was small compared to the rest of the animal, the eyes were large, and the snout was long. The teeth were small and pointed and not like the dagger-shaped teeth of typical flesh-eating dinosaurs.

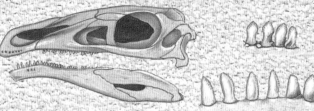

Erlikosaurus skull

Each of the three fingers of therizinosaurs bore a huge claw. The claw (below) was long and narrow, but straight rather than hooked as it is in flesh-eating dinosaurs.

Alxasaurus claw

The arms of Alxasaurus are quite long, while those of Therizinosaurus, its larger descendant, are extraordinarily large and perhaps longer even than the hind limbs. The upper arm bone, the humerus, bears an enormous bony crest for the attachment of a huge chest, or pectoral muscles, and shows that therizinosaurs had extremely powerful arms.

The hips of therizinosaurs are massive and the legs relatively short and sturdy (right). The feet are quite broad and therizinosaurs probably walked in a flat-footed fashion rather than running on their toes.

139

Horned Dinos

Ceratops, the "horn-faced" dinosaurs, were so common throughout the Late Cretaceous that they have been called "the sheep of the Dinosaur Age." Early forms, such as Protoceratops, were not much larger than sheep, while later Ceratops, such as Triceratops, were often as big as a rhinoceros, or even larger. These plant-eaters had strange crests and horns around the skull and parrot-like beaks that were used to chomp up tough plant material. Even though many Ceratops have already been unearthed and described, new ones including Einiosaurus, discovered recently in Montana, continue to be found.

Triceratops

Immense bone beds consisting of the jumbled up skeletons of hundreds and sometimes thousands of Ceratops have been found in North America. It seems likely that these animals were killed by a catastrophe, such as a huge flood.

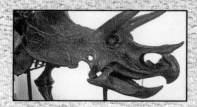

FEEDING

Ceratops had a toothless beak, which they used to bite off plant twigs and leaves. Tougher material was sliced up in a scissor-like fashion by teeth at the back of the mouth (left).

Ceratops teeth

HORNS AND FRILLS

Ceratops are distinguished by an astonishing variety of horns as well as a selection of frills that were also edged in horns (below). The full display was only developed when they reached full adulthood.

Chasmosaurus

Styracosaurus

Ceratops were almost certainly preyed upon by theropod hunters. If they were threatened by a predator, herds of Einiosaurus could protect their young by bunching together with the adults facing out on the outside (below). This is similar to the behavior of modern-day Musk Oxen from the Arctic.

WHAT'S IN A NAME

Each species of dinosaur has a scientific name composed of two words, written in Latin or Greek. The name usually describes a predominant feature of the animal. For example, Einiosaurus procurvicornis means "buffalo lizard with a forward-curving horn."

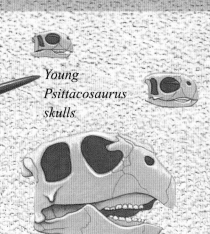

Young Psittacosaurus skulls

YOUNGSTERS

In some dinosaurs, such as Psittacosaurus, an early Ceratops, there are fossil remains for all stages of growth, from the very young to the fully-grown adult (above). Like human babies, young dinosaurs had skulls with large eyes and separate bones to allow for lots of growth. These separate bones would fuse together to form a single skull in later life (left).

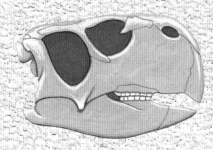

It seems likely that the horns were used for display purposes and also for fighting. This points to the existence of hierarchies within herds. The horns may have been used to gouge and stab attacking animals or even other members of the same herd. They may even have become interlocked (right), as can happen with the horns of modern antelope and deer.

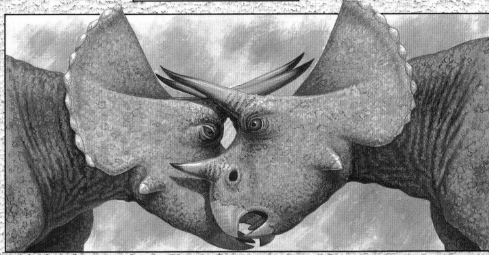

141

Giant DINOS

There is probably a limit to the size to which animals can grow, and it seems likely that sauropods got close to that limit. Seismosaurus, from the Late Jurassic and found in New Mexico, was about 150 feet (46 m) from nose to tail. It was probably the longest sauropod. Ultrasauros, found in Utah, was 50 feet (15 m) or more in height and was certainly the tallest. In terms of sheer size, however, a recent discovery from Argentina, Argentinosaurus, currently holds the record for the biggest dinosaur, with an estimated weight of up to 100 tons. It takes time to grow to such a large size and animals such as these probably lived for 50 to 100 years, or perhaps even centuries!

LATEST FINDS

The latest discoveries of sauropods come from Madagascar. The most important find is an almost complete titanosaur. So far, fossils of this important but poorly known group of Late Cretaceous sauropods have been very fragmentary.

Ultrasauros
This huge plant-eater (right) was more than 82 feet (25 m) long and weighed up to 50 tons (45 t). In relation to the rest of its body, it had a small head that sat on top of a towering giraffe-like neck.

Supersaurus
Weighing about the same as Ultrasauros (see left), this dinosaur (below) had a neck that was 32 feet (10 m) long, allowing it to pick leaves from very tall trees.

SAUROPODS GET A GRIP

The hand of sauropods had a claw on the thumb. Scientists have often wondered what its function was. New findings show that it was probably used to grip trees as they stood on hind limbs to reach high into the foliage for the most succulent leaves (right).

Sauropods are long thought to have eaten stones and kept them in their gizzard to help grind up food. New discoveries on the Isle of Wight in Britain show that this was indeed the case.

Argentinosaurus
At three times the size of Seismosaurus, and with an estimated length of more than 100 feet (30 m), Argentinosaurus may have been the largest land animal ever. Each of its fossil vertebrae alone weighs a ton!

AMARGASAURUS

Amargasaurus (below), a sauropod from the Early Cretaceous and discovered in Argentina, was 33 feet (10 m) long and had pairs of spines that ran down its neck and along its back. These spines supported an extraordinary sail, which might have been used for display or perhaps for defense, preventing theropods from biting through Amargasaurus' neck.

Seismosaurus, the ground shaker from the Late Jurassic, is probably the longest dinosaur that ever lived. To compensate for its extreme length, it may have had rather short legs to give it stability.

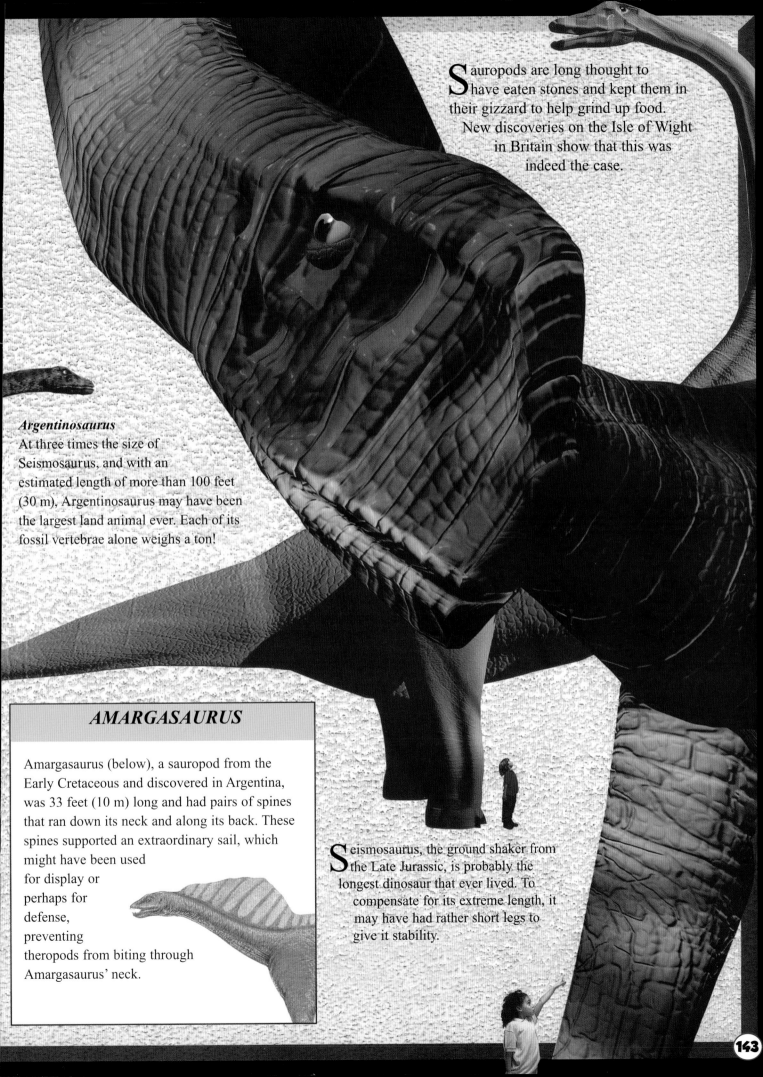

Birds that FILL the GAP

Archaeopteryx, the most primitive known bird (pterosaurs were not birds), has lots of dinosaur characteristics and shows that birds evolved from theropods (see pages 136-137). Until recently, Archaeopteryx and one or two other species, such as Hesperornis, were the only known birds from the age of the dinosaurs. There was little evidence indicating exactly when and how birds evolved. Fortunately, the last ten years have seen some spectacular discoveries. These include fossils from China with feathers preserved intact (below), and the dinosaur-like bird Mononykus (right) found in Mongolia. These show how the bird wing evolved and how other important features such as perching developed.

FLOCK OF BIRDS FROM SPAIN

Important discoveries of early birds have recently been made in Spain. The skeletons of Nogueornis and Concornis have features which show that although these birds were around only a few million years after Archaeopteryx, they had well-developed wings and were strong fliers. Special adaptations in the foot also allowed them to perch on branches like modern birds.

Skeleton of Protoavis

Skeleton of Mononykus

Called Mononykus (left), this animal from the Late Cretaceous, was about the size of a turkey. It had a skull very similar to a bird's, but a pelvis not unlike that of a dinosaur. The most distinctive features were its short, stubby forelimbs, each with a single claw. Scientists cannot agree whether this was a flightless bird or a peculiar burrowing dinosaur that used its forelimbs to dig into the ground.

ARCHAEOPTERYX

The relationship between dinosaurs and modern birds first came to light with the discovery of a feathered dinosaur in 1861. This animal was named Archaeopteryx ("ancient wing"—right and below), and was clearly an early bird, but it had one or two strange features. Although the head is fairly birdlike, its jaw was lined with small teeth, whereas birds have beaks or bills. Archaeopteryx also had a long, bony tail with feathers growing along it. In contrast, modern birds' tail feathers all grow from the base of a stubby tail.

Fossil remains of Archaeopteryx

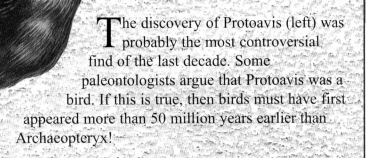

The discovery of Protoavis (left) was probably the most controversial find of the last decade. Some paleontologists argue that Protoavis was a bird. If this is true, then birds must have first appeared more than 50 million years earlier than Archaeopteryx!

AGE OF THE BIRDS

The diagram (below) shows when various birds lived. The examples include Archaeopteryx (1), Mononykus (2), Enantiornithines (3), Patagopteryx (4), Hesperornithiforms (5), Ichthyornithiforms (6), and modern birds (7).

CONFUCIUSORNIS

In the last few years, many examples of Confuciusornis (right) have been found in China dating from the Early Cretaceous. It was well-designed for tree climbing, but its wings were far too short to make Confuciusornis a very good flier.

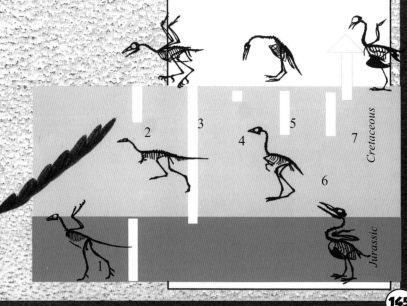

Extinction Theories

Death is inevitable for all individuals, and species are always disappearing—this is called "background extinction." Sometimes, entire groups can disappear. This occurred at the end of the Cretaceous (65 million years ago), when the dinosaurs died out, along with many other types of animal and plant life. Such events are called "mass extinctions." As well as being bad news for some groups, it can be good news for others, allowing them to arise from obscurity and take over new habitats. Many theories exist as to what caused mass extinctions. These can range from the serious, such as volcanic activity and a meteorite impact, to the bizarre, such as a space borne plague or even alien hunters!

VOLCANIC ERUPTIONS

Huge volcanic eruptions (above) are known to have taken place at the same time as mass extinctions, including the one at the end of the Cretaceous. These eruptions may have led to acid rain on a global scale, a mini-ice age, and then an abrupt swing into a greenhouse type climate. This could have killed off the dinosaurs.

Many other groups, including the pterosaurs, marine reptiles, such as plesiosaurs and mosasaurs (see page 127), other forms of oceanic life, which include the ammonites (right), and some kinds of plants all died out together with the dinosaurs.

TARGET EARTH

The impact of a large meteorite may have caused a global catastrophe big enough to wipe out the dinosaurs. But where's the crater? A huge bowl-shaped structure more than 113 miles (182 km) across has been found near Chixulub, (the "Devil's Tail", below) in Mexico. It was formed 65 million years ago, exactly when the dinosaurs died out!

Mexico

Meteorite impacts are very common—the Earth is constantly being hit by small bits of debris from space. However, large impacts occur less frequently. Meteor Crater in Arizona (left) was caused by a meteor that hit the Earth only 50,000 years ago.

A mass extinction at the end of the Triassic may have allowed the dinosaurs to rise to prominence, while another at the end of the Cretaceous finished off the dinosaurs and gave mammals their chance to evolve (below).

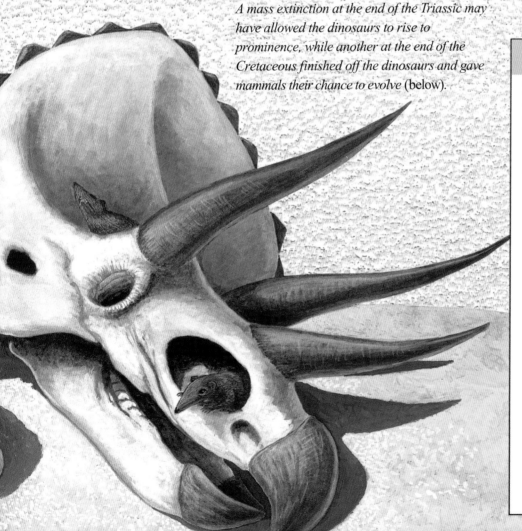

ARE THEY EXTINCT?

In many ways, it can be said that not all of the dinosaurs were killed off by the mysterious mass extinction 65 million years ago. Their descendants, the birds, can be found on every continent of the world. These successful animals range in size from tiny hummingbirds to eagles, penguins, and enormous ostriches (below).

SHARKS GLOSSARY

Barbel
A fingerlike feeler on the side of a shark's nose or jaw.

Basking
When referring to sharks, this means swimming near the surface of the water.

Camouflage
Being colored or shaped to blend in easily with your surroundings.

Cartilage
A tough, flexible material that makes up the skeleton of sharks, skates, and rays.

Denticles
The thornlike points covering a shark's body.

Dorsal
On the back, as in a dorsal fin.

Filter Feeder
Any sea creature that feeds by removing tiny plants and animals from sea water.

Fin
Part of a fish that helps it to move, steer, or balance in water.

Fossil
The remains of living things that lived millions of years ago.

Gills
Tiny featherlike parts inside the throat or behind the skull of a fish or shark that take oxygen out of water.

Lateral Line
A special line along a shark's sides that helps it to detect movements in the water.

Pectoral
On the chest, as in a pectoral fin.

Plankton
Small, often almost invisible, plants and animals that drift in the seas.

Pore
A tiny hole in the skin.

Pup
A newborn shark.

Spine
A hard, pointed part on a fish. It is often sharp and may contain poison.

Streamlined
Having a smooth shape that will move easily through air or water.

Swim Bladder
A special balloon-like bag filled with gas inside the body of bony fish. It helps fish to float in the water.

HUMAN BODY GLOSSARY

Anatomy
The study of the structure of living things, as opposed to physiology, which is the study of how living things function. Human anatomy is the study of the structure of the human body.

Artery
A large tube or vessel that carries blood away from the heart to the organs and tissues. Most arteries contain bright red, high-oxygen blood. The pulmonary arteries carry dark, low-oxygen blood from the heart to the lungs.

Capillaries
A network of very small tubes or vessels, finer than hairs. The blood inside them can pass its oxygen and nutrients to the surrounding tissues.

Cartilage
A smooth, rubbery, or gristly substance. It covers the ends of bones in joints and forms a flexible framework for parts such as the nose, ears, and voice box.

Cell
The basic building block of living things. There are billions of cells in the human body, of many different kinds—muscle cells, bone cells, blood cells, and so on. An average cell is one-hundredth of a millimeter across and can only be seen with a microscope.

DNA
A molecule that contains all the genetic information passed from parents to offspring and tells the body how to develop. DNA is a very long molecule, consisting of two twisted strands.

Enzyme
A body substance that speeds up a chemical reaction. The body has hundreds of enzymes. Digestive enzymes speed up the process of digestion, dissolving food into smaller pieces.

Gland
A body part that makes a substance, usually a liquid, that has a specific use. The pancreas gland makes digestive juices to attack food. Endocrine glands, such as the adrenals, make hormones. In a woman, mammary glands produce milk, which is all the food a new baby needs for the first months of its life.

Hormone
A body substance that acts as a "chemical messenger." Made in an endocrine gland, it circulates in the blood, and affects the way certain cells and organs work.

Ligament
A strong, slightly stretchy, strap-like part around a joint. Several ligaments hold the bones of a joint together, strengthening and stabilizing the joint.

Membrane
The thin covering or "skin" over a body cell or larger part.

Muscle
A body part specialized to contract, or get shorter. Muscles make all the movements in the body, both inside the body (stomach and intestines), and movements of the whole body.

Nerve
A body part specialized to carry messages in the form of tiny electrical signals or pulses. Nerves control and coordinate most body processes.

Organ
A major body part that has a distinct function, such as the heart, lung, or brain.

Tendon
A strong, rope-like part at each end of a muscle. The tendon connects the muscle itself to the bone that it pulls.

Tissue
A group of similar cells that all do the same job, such as muscle tissue or nerve tissue. Collections of different tissues make up the major body parts, known as organs.

Vein
A large tube or vessel that carries blood back to the heart from the organs and tissues. Most veins contain dark, low-oxygen blood. The pulmonary veins carry bright red, high-oxygen blood from the lungs to the heart.

INSECTS GLOSSARY

Abdomen
The rear part of an insect's body, which contains the internal organs for digestion, excretion, and reproduction.

Camouflage
Patterns, colors, and shapes on an insect's body that make it blend in with the background, so it is difficult to see.

Carnivore
An animal that eats only other animals.

Elytra
The hard pair of wing-cases that cover a beetle's wings.

Entomology
The study of insects.

Enzymes
Special chemicals made by living things that help chemical reactions to take place.

Exoskeleton
The hard outside skin on the bodies of insects and their relatives, which protects and supports their soft parts.

Gall
A swelling made by some plants when attacked by insects such as tiny wasps.

Herbivore
An animal that eats only plants.

Labium
The part that forms the top of an insect's mouth, or upper "lip."

Labrum
The part which forms the bottom of an insect's mouth, or lower "lip."

Larvae
The young grubs, maggots, or caterpillars of insects that undergo complete metamorphosis.

Mandibles
A pair of jaws which bite together in front of an insect's mouth.

Maxilla
A second pair of jaws, behind the mandibles, in front of an insect's mouth.

Metamorphosis
The transformation of a young insect to an adult.

Migration
A journey undertaken by an animal to avoid bad conditions such as extreme cold or heat, or lack of food or water.

Nymph
The young of insects that undergo incomplete metamorphosis.

Parasite
A plant or animal which eats another plant or animal while it is still alive.

Pheromones
Special scents given off by animals at certain times, such as in the breeding season, to communicate with others.

Polarized light
Light waves that vibrate in the same plane. Humans cannot tell it from unpolarized light that vibrates in all planes, but many insects can.

Predator
An animal that kills and eats other animals for food.

Prey
An animal hunted for food.

Proboscis
A tube formed from the mouthparts of insects, such as flies and butterflies, used to suck up liquid food.

Pupa
The chrysalis stage, between the larva and the adult, in insects that undergo complete metamorphosis.

Rostrum
The rigid, needle-like tube formed from the mouth parts of insects, such as bugs, used to pierce and suck up juices.

Spiracles
Holes in the exoskeleton that lead to breathing tubes within an insect's body.

Thorax
The middle part of an insect's body, to which the wings and legs are attached.

Trachea
Breathing tubes that branch inside an insect's body, carrying air to the muscles and organs.

DINOSAUR GLOSSARY

Ammonites
Coiled shellfish that are closely related to squid and octopus. They died out at the end of the Cretaceous Period.

Ankylosaurs
Four-legged, plant-eating dinosaurs that had heavily armored bodies and sometimes massive clubs on the end of their tails. They include Ankylosaurus and Euoplocephalus.

Ceratopsians
Four-legged, plant-eating dinosaurs that often had beaks and horns. They include Triceratops and Protoceratops.

Coprolites
The fossilized remains of dinosaur dung.

Cretaceous Period
A period in the Earth's history that lasted from 135 million years ago until the extinction of the dinosaurs, some 65 million years ago.

Dinosaurs
Taken from the Greek for "terrible, or fearfully great lizard," the term refers to a group of land-living reptiles that flourished between 225 million years ago and 65 million years ago when they became extinct.

DNA
Short for "deoxyribonucleic acid," this is the recipe that cells use to build living organisms. It is stored in the cell's nucleus.

Embryos
The very early stages of a young animal before it has been born or hatched.

Extinction
When a whole group or species of animals or plants dies out.

Fossils
The preserved remains of something that was once alive. They can be formed in a number of ways, including burial and the chemical change of the body parts into minerals.

Gondwana
The large and ancient southern continent that was made up of what is now South America, Africa, India, Antarctica, and Australia.

Gymnosperms
A group of plants that do not produce flowers. The last remaining example of these once widespread plants is the ginkgo.

Hadrosaurs
Plant-eating dinosaurs that often had duckbill-shaped mouths and a crest on the top of their heads. They include Edmotosaurus and Maiasaura.

Hibernation
During winter, many animals hibernate to conserve energy through the cold months. This involves slowing down the body's processes and spending the time in a sleep-like state.

Ichthyosaurs
A group of sea-living reptiles from the age of the dinosaurs. These swimming animals had very streamlined, fish-shaped bodies.

Jurassic Period
A period in the Earth's history that lasted from 200 million years ago to 135 million years ago.

Laurasia
The large and ancient northern continent that was made up of what is now North America, Europe, and Asia.

Migration
A movement from one place to another. Animals generally migrate to avoid the cold of winter or to search for food.

Timeline

PERMIAN
290 MYA

TRIASSIC
245 MYA

JURASSIC
225 MYA

Paleoichnology
The study of dinosaur footprints.

Paleontologists
People who study fossils.

Permian Period
A period in the Earth's history, lasting from 290 million years ago to 245 million years ago.

Plesiosaurs
Although not actually dinosaurs, these flesh-eating sea creatures were around at the same time.

Pterosaurs
Flying reptiles from the dinosaur era. They were distantly related to dinosaurs.

Sauropods
Large, plant-eating, four-legged dinosaurs. They included Seismosaurus and Diplodocus.

Tertiary Period
A period in the Earth's history that lasted from 65 million years ago to 1.8 million years ago.

Tethys Sea
A body of water that separated Laurasia and Gondwana.

Therizinosaurs
Plant-eating dinosaurs with long arms and huge claws. They include Alxasaurus and Therizinosaurus.

Theropods
A range of hunting dinosaurs that usually stood on two legs. They include Tyrannosaurus and Eoraptor.

Titanosaurs
Very large sauropods.

Triassic Period
A period in the Earth's history that lasted from 245 million years ago to 200 million years ago. The dinosaurs first appeared toward the end of this period.

Tyrannosaurs
Large, two-legged, flesh-eating dinosaurs. They include Tyrannosaurus and Albertosaurus.

Ultrasauros   Confuciusornis   Afrovenator

Extinction of the dinosaurs

CRETACEOUS
135 MYA

TERTIARY
65 MYA

INDEX

AIDS 83
abdomen 98, 99, 121
acacia 20
Aconcagua, Mount 24
Afghanistan 32
Africa 17, 18, 20, 28, 29
Afrovenator 137
Alaska 22, 37
Albania 26
Albertosaurus 135, 149
Algeria 28
Alps 16, 27
Alxasaurus 138-139, 143
Amargasaurus 143
Amazon, river 14, 24
amber 128
America 17, 18, 20, 21, 22, 24, 25
    Central 12, 24
    North 20, 21, 22, 23
    South 17, 18, 20, 24, 25
American Samoa 35
ammonites 146, 148
Andes 16, 24
Andorra 26
angel shark 48, 56, 57
Angola 28
ankylosaurs 129, 148
ant 99, 104, 118-119
Antarctica 17, 19, 36
Antarctic Peninsula 36
antennae 98, 104, 105
antibodies 82
Arabian Desert 18
Archaeopteryx 144, 145
Arctic 19, 37
Arctic Circle 27, 30, 37
Arctic fox 19
Arctic Ocean 19, 37
Argentina 25, 36
Argentinosaurus 142, 143
Armenia 27
Arteries 78
Asia 14, 17, 20, 27, 30, 31, 32, 33
    northern 30, 31
    Southeast 14, 32, 33
Atacama Desert 18
Atlantic Ocean 15, 24, 29
Atlas Mountains 16, 29
Australia 15, 21, 34, 35
Australian desert 18
Austria 26
Automatic nerve network 88
Azerbaijan 27

Bahrain 29
Baikal, lake 14
Balance, sense of 84
Baltic Sea 17
Bangladesh 32, 33
baobab 20
barbels 48, 65
Baryonyx 137
basking shark 42, 46, 56, 57, 64
bees and wasps 99, 101, 102, 104, 105, 106, 107, 116-117
beetles 99, 102, 103, 106, 107, 109, 110-111
Belarus 26
Belgium 26
Belize 24
Benin 26
Bhutan 32
big horn sheep 16
birds 126, 144-145, 147
Black Sea 15
bladder 80
blood 78, 82
blood-suckers 108, 112
blood groups 79
blood vessels 78, 82
blue shark 43, 46
body-building 92
body design 98-99
body temperature 114
Bolivia 25
bones 72, 73
Bosnia-Herzegovinia 26
Botswana 28
brain 71, 88-89, 99, 104
Brazil 25
breathing 74-75, 88
Brunei 33
bugs 100, 108-109
Bulgaria 26
bull shark 47, 55
butterflies and moths 99, 100, 101, 103, 105, 107, 114-115
Burkino Faso 28
Burundi 28

Cambodia 33
camel 18
Cameroon 28
camouflage 57, 59, 67, 98, 108, 115, 121
Canada 25, 26, 37
Cape Horn 24

capillaries 75, 78
Carcharodontosaurus 136
Caribbean Sea 24
carnivores 110, 121
Carpathian Mountains 27
carpet shark 46, 56, 59
cartilage 73
Caspian Sea 14
caterpillars 101, 102, 107, 115
cells 70
Central African Republic 28
ceratopsians 131, 133, 140-145, 148
Cerebral cortex 88
Chad 28
champsosaurs 138
Chang jiang (Yangtze) 30
Chasmosaurus 140
Chile 25, 36
China 31
cicadas 109
Circulatory system 71, 78-79
clams 15
classification of insects 120
climate 16, 18
coastline 15
Cochlea 84
cockroaches 98, 100
coelurosaurs 135
Colombia 24
Colorado River 22
Concornis 144
Confuciusornis 145
Congo 28, 29
Congo River 29
coniferous evergreens 17
continental shelf 15
continental slope 15
continents 21, 22, 24, 27, 29, 30, 26, 37
cookie cutter shark 58, 24
coral 15
    atolls 35
    reefs 15
Costa Rica 24
Côte D'Ivoire 28
cottonwood 20
crabs 15
craters 147
Cretaceous Period 132, 138, 140, 142, 144, 145, 146, 147, 148
crickets 98, 105
Croatia 26
crust 16

INDEX

Cuba 24
cuts and wounds 83
Cyprus 26
Czech Republic 27

DNA 91, 128, 129, 148
Danube River 27
Darling River 35
Daspletosaurus 135
deathwatch beetle 111
deciduous forests 16
deer 17
deforestation 17
deltas 14
Deltadromeus 136
Denmark 26
denticles 44, 65
deserts 18, 20, 24, 35
diaphragm 71, 74
digestive system 76-77, 88
Dinosaur Freeway
Diplodocus 149 130, 131
disease 87
disease carriers 108, 112
Djibouti 29
dogfish 42, 52, 58
Dominica 25
Dominican Republic 24
dragonflies 98, 100, 104
dung 148
dwarf shark 42

ear 84
Easter Island 35
Ecuador 24
Edmontosaurus 148
Egypt 28
eggs 52, 53, 100, 101, 128, 132, 133
egg teeth 133
Einiosaurus 140-145
El Salvador 24
elytra 111, 121
embryos 132-133, 148
enantiornithines 145
energy 76
entomology 115, 121
enzyme 77, 80, 110, 113, 121
Eoraptor 136, 149
Equatorial Guinea 28
Eritrea 28
Erlikosaurus 139
esophagus 76-77
Estonia 27

Ethiopia 28
Ethiopian Highlands 29
Europe 17, 26, 27, 35
European plain 27
Eustachian tube 84 Everest, Mount 33
exercise 75
exoskeleton 98, 99, 121
extinction 146, 148
eyes 85, 98, 104

Falkland Islands 36
farming 20, 27
feeding 51
feeding frenzy 51
feeding habits 108, 109, 112
Fennec fox 18
Fetus 91
Fiji 35
filter feeding 51
Finland 35
fins 45, 65
fjord 35
fleas 103
flies 99, 102, 104, 112-113
flight 101, 111, 112
flowers 126
flowering plants 16
forests 16, 17, 20, 22, 30
   boreal 17, 22, 30
   coniferous 16, 22
   rain 17
   temperate 17
   tropical cloud 16
fossils 59, 67, 128, 129, 142, 145, 148
France 26
French Guiana 25
freshwater 14
frilled shark 58
Fuji, Mount 30
F.Y.R.O.M. (Macedonia) 26

Gabon 28
Gambia, The 28
Ganges River 14, 33
Georgia 27
Germ 82
Germany 26
Ghana 28
Giganotosaurus 137
gills 49, 65
gingkoes 138, 148

giraffe 20
glaciers 19
glands 81
glowworms (fireflies) 110
Gobi Desert 18, 30
goblin shark 59
Gondwana 126, 148, 149
grasses 16, 19, 20
grasslands 16, 20, 29, 30
Great Dividing Range 16, 35
Great Lakes, The 14
Great Rift Valley 29
Great Sandy Desert 35
Great Victoria Desert 35
great white shark 43, 49, 52, 55
Greece 26
green dogfish 58
Greenland 19, 32, 33, 37
Grenada 25
grey reef shark 43, 55, 64
growth 81
grubs 101, 106, 110, 112, 116, 118
Guatemala 24

Guinea 28
Guinea-Bissau 28
Guyana 25
gymnosperms 138, 148

HIV 83
hadrosaurs 129, 148
Haiti 24
hammerhead shark 43, 46, 48, 53, 58
Hawaiian Islands 22
hills 16
hearing 84, 86, 104

155

INDEX

heart 71, 78-79
heartbeat 79
herbivores 110, 121
Hesperornis 144
hesperornithiforms 145
hibernation 148
Himalayan Plateau 33
Himalayas 16, 33
Honduras 24
hormones 80-81
horn shark 46, 50, 53, 58
horns 140-141
Hungary 26

ice 16, 19, 22, 36, 37
ice caps 126
ice sheets 19, 22, 36, 37
icebergs 19
Iceland 26, 27
ichthyornithiforms 145
ichthyosaurs 127, 148
Immune system 82
insects 128
India 32
Indian Ocean 15, 33
Indonesia 33
intestines 71, 76-77, 80
Iran 29
Iraq 29
Israel 28
Italy 26

Jamaica 24
Japan 30, 31
Java 33
joints 72
Jordan 28
Jurassic Park 128
Jurassic Period 126, 142, 143, 145, 149

Kalahari Desert 18
Karakumy Desert 18
Kazakhstan 30
Kenya 28
kidneys 71, 80
Kilimanjaro, Mount 29
Kiribati 35
Kosciusko, Mount 35
Kuwait 29
Kyrgyzstan 30

labium 121

labrum 121
lakes 14
lantern shark 42, 58
Laos 33
larvae 100, 101, 111, 121
lateral line 48, 65
Latvia 36
Laurasia 126, 149
Lebanon 28
legs 98, 102, 103
lemon shark 43, 53
leopard shark 42, 43, 49
Lesotho 28
Liberia 28
Libya 28
Liechtenstein 26
life cycles 100, 101
ligament 73
Lithuania 26
liver 71, 80
livestock 20
lizards 126, 128
locusts 100
lungs 71, 74-75
Luxembourg 26
lymphatic system 82

Macedonia see F.Y.R.O.M. (Macedonia)
Madagascar 29
maggots 102
Maiasaura 148
mako shark 43, 45
Malawi 28
Malaysia 33
Maldives 33
Mali 28
Malta 26
mammals 126, 127, 128, 147
mandibles 121
manta ray 61
Mariana Trench 14
Martinique 35
Masai Mara 29
mating 52, 105, 109, 116, 119
Mauritania 28
Mauritius 29
maxilla 121
McKinley, Mount 22
meandering rivers 14
medicine 62
Mediterranean Sea 15, 27
megalodon 60

megamouth shark 50, 59
Megazostrodon 127
Mekong River 14, 33
membrane 70
menstrual cycle 81, 90
mermaid's purse 54
metamorphosis 100, 101, 113, 121
meteorites 146, 147
Mexico 24
migration 103, 121, 130, 149
mimicry 114
Mississippi River 14, 22
Missouri River 22
Mojave Desert 18
Moldova 26
Monaco 26
Mongolia 31
Monolophosaurus 137
Mononykus 144, 145
Morocco 28
mosquitoes 112
mosses 19
mountain avens 16
mountains 16, 18, 19, 22, 24, 29, 36, 37
mouthparts 98
Mozambique 28, 29
Murray River 35

INDEX

muscles 92-93, 128, 129
musk ox 19
Myanmar 32

Namibia 28
Nauru 35
Nepal 32
nerve impulses 84-85, 86
nervous system 71
nests 116, 118, 129, 132, 133
Netherlands, The 26
New Caledonia 35
New Mexico 22
New Zealand 35
Nicaragua 24
Niger 28
Niger River 29
Nigeria 28
Nile River 14, 29
Nogueornis 144
North Korea 31
North Pole 19, 37
Norway 26, 37
nose 86
nurse shark 46, 48, 56, 59
nutrition 76
nymphs 100, 121

ocean floor 15
oceans 15, 21
Oman 29
organs 70, 71
ornithomimosaurs 129
ovaries 81
Oviraptor 132-133

Pacific Ocean 16, 17, 33, 35
pack-ice 36
Pakistan 31, 33
paleontoligists 128, 129, 132-133, 135, 138-139, 145, 149
Pampas 20
Panama 24
Pancreas 80-81
Papua New Guinea 34
Paraguay 35
Parana River 14
parthenogenesis 109
Patagonia Desert 18
Patagopteryx 145
Pelecanimimus 129
pelvises 134, 135, 144
permafrost 19

Permian Period 149
Peru 35
pests 108, 110, 111
pheromones 105, 118, 121
Philippines 33
pilot fish 49
pine martens 17
pingoes 19
pla beuk 14
placenta 91
plankton 15, 50, 60, 65
plants 126, 138, 146
platelets 78
plates 16
plesiosaurs 127, 146, 149
poisonous insects 106-107, 114
Poland 26
Poles 126
porbeagle shark 43
pores 48
Port Jackson shark 58
Portugal 26
prairie dogs 20
prairies 20, 22
pregnancy 91
prehistoric insects 98
proboscis 112, 118
Protoavis 144, 145
Protoceratops 133, 140, 148
Psittacosaurus 138, 141
pterosaurs 126, 127, 130, 130, 133, 146, 149
pulse 79
pupa 101, 118, 121
pups 52, 53, 67
Pyrenees 27

Qatar 29

rainfall 17, 18
rainforest 17, 24, 29, 35
rain-shadow deserts 18
rays 60, 61
Red Sea 29
remora 49
reproductive system 71, 81, 90-91
Republic of Ireland 26
respiratory system 71, 74-75
Rhine River 27
Rhône River 27
rice 33
Ring of Fire 30
Rio Grande 22

rivers 14
Rockies, The 16
Romania 26
rostrum 108, 121
Russian Federation 26, 27, 30, 31
Rwanda 26

St. Helens, Mount 16
St. Lawrence River 22
Sahara Desert 18
Sahel 20, 29
sand dunes 18
Sao Tome and Principe 28
Saudi Arabia 29
sauropods 130, 131, 132, 139, 142-143, 149
savanna 20, 29
sawfish 61
scarab beetle 111
scrub 24, 27, 33, 35
seas 15, 21
Seismosaurus 142, 143, 149
Senegal 28
sense organs 104-105, 118
sensory system 84-85, 87
sex glands 81
shagreen 63
shark attacks 55
shrubs 20
Siamotyrannus 135
Siberia 17
Sierra Leone 28
sight 84-85, 86
Singapore 33
Sinornithoides 137
Sinraptor 137
skates 61
skeleton 44, 72, 73, 128, 129, 134, 138, 144
skin 86-87, 128, 129
skulls 129, 134, 136, 139, 145
sleeper shark 47
Slovakia 26
Slovenia 26
smell, sense of 84
snow 16
soft tissues 128-129
Solomon Islands 34
Somalia 29
South Africa, Republic of 28
South Korea 31
South Georgia 36
South Orkney 36

INDEX

South Pacific Ocean 15
South Pole 19
Spain 26
spiders 99
spiracles 98, 99, 119, 121
squirrels 17
Sri Lanka 33
stag beetle 101, 121
steppes 20, 30
stings 106, 107, 116
stingrays 15, 61
stomach 71, 76, 80
streamlining 44, 45, 65
Styracosaurus 140
Sudan 28
Sumatra 33
Surinam 25
swamps 14
Swaziland 28
sweat 87
Sweden 26
swell shark 46, 53, 57, 59
swim bladder 44, 65
Switzerland 26
Syria 28, 29

taiga 17, 29
Taiwan 30
Tajikistan 30
Tanzania 28, 29
Tarbosaurus 135
taste buds 86
taste sensors 104
Taymyr Peninsula 19
teeth 50, 76, 129, 138, 139, 140, 145
tendons 81, 90, 92
termites 118, 119
Tertiary Period 149
Thailand 33
therizinosaurs 132, 138, 139, 149
theropods 131, 136-137, 139, 141, 143, 144, 149
thorax 98, 99, 102, 103, 112, 121
tigers 17
tiger shark 51, 55
tissues 70
titanosaurs 132, 142, 149
Togo 28
Tonga 35
toucans 17
trachea 74
tracks 130-131
traditional medicine 71, 82, 84, 89

Transantarctic Mountains 36
trees 20
Triassic Period 147, 149
Triceratops 131, 140, 148
Trinidad and Tobago 24
tundra 21, 28, 30, 37
Tunisia 28
Turkey 26
Turkmenistan 30
Tuvalu 35
Twelve Apostles 15
Tyrannosaurus 134-135, 136, 137, 149

Uganda 28
Ukraine 26
Ultrasauros 142
United Arab Emirates 29
United Kingdom 26
United States of America 22, 23
Urals 16, 27
urinary system 80
Uruguay 35
uterus 81
Uzbekistan 30

vaccination 82
Vanuatu 35
Vatican City 26
veins 78
Venezuela 24, 35
Vietnam 33
Vinson Massif 36
Vistula, river 27
volcanoes 16, 30, 146
Volga, river 27, 30

wastes 78, 80
water 15
water cycle 14
water insects 103, 108, 109
waterfalls 14
Western Sahara 28
Western Samoa 35
whales 15
whale shark 42, 43, 56, 57
wings 98, 99, 102, 108, 111, 112, 114
wobbegong shark 48, 59
Wrangel Island 37
Wyoming 22

X-rays and scans 72, 73

Yangtze River 30
Yellowstone National Park 22
Yemen 29
Yucatan Peninsula 24
Yugoslavia 26
Yukon River 22

Zaire 28
Zambia 28
zebra 20
Zimbabwe 26

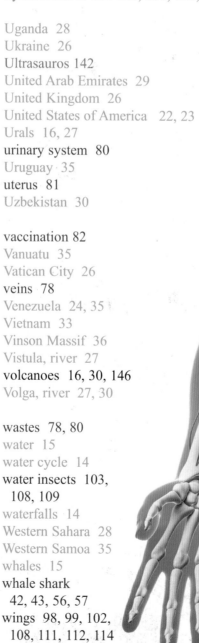